✓ **Y0-ARX-290**

DISCARDED

Errata

1. pg. 42 last line should be: non-substitutionally false
2. pg. 15 line 4 delete left parenthesis: (Unless
3. pg. 118 line 15 should be: Recall Kant's remark
4. pg. 149 10 lines up from bottom should be: intensional
5. pg. 168 first line should be: (∃F)(Fb)

DISCARDED

Existence
and the
Particular Quantifier

 **PHILOSOPHICAL MONOGRAPHS**

Joseph Margolis, General Editor
Temple University
Philadelphia, Pa. 19122

Assistants to the Editor

Tom L. Beauchamp
Georgetown
Donald Harward
Delaware

Judith Tormey
Temple

Editorial Advisory Board

Kurt Baier
Pittsburgh

Stephen F. Barker
Johns Hopkins

Panayot Butchvarov
Iowa

Norman Care
Oberlin

V. C. Chappell
Massachusetts, Amherst

Keith Donnellan
California, Los Angeles

Jerry Fodor
MIT

Donald Gustafson
Cincinnati

Michael Lockwood
Oxford

Ernan McMullin
Notre Dame

Harold Morick
SUNY, Albany

Terence Penelhum
Calgary

Jerome Shaffer
Connecticut

H. S. Thayer
CUNY

Zak Van Straaten
Witwatersrand

Richard Wasserstrom
California, Los Angeles

Existence
and the
Particular Quantifier

Alex Orenstein

Upsala College
Library
East Orange, N. J. 07019

Philosophical Monographs
Second Annual Series

Temple University Press
Philadelphia

Library of Congress Cataloging in Publication Data

Orenstein, Alex.
 Existence and the particular quantifier.

 (Philosophical monographs)
 Bibliography: p.
 Includes index.
 1. Logic. 2. Ontology. 3. Predicate calculus.
I. Title. II. Series: Philosophical monographs
(Philadelphia, 1978-   )
BC57.073   160   78-14515
ISBN 0-87722-130-8
ISBN 0-87722-126-X pbk.

Temple University Press, Philadelphia 19122
©1978 by Temple University. All rights reversed
Published 1978
Printed in the United States of America
ISSN 0363-8243

160435

to the memory of
James Gordon Clapp

Acknowledgments

Most of the material in this monograph dates back to my doctoral dissertation (New York University, 1972) and I still am indebted to Raziel Abelson, Henry Hiz, and Milton K. Munitz, my advisors at that time. Since then, however, my thoughts on this subject have undergone many metamorphoses. Earlier drafts of parts of the monograph have been given as lectures at the Graduate Center of the City University of New York (1973) and at a conference on Ontology at the Biltmore Hotel, New York (1975). Some material has even appeared in print. Thus in a recent book on W. V. Quine (Twayne Publishers, 1977), I adopted several ideas to interpret Quine's views on ontological commitment and quantification. The papers "Strawson, Frege, and Hilbert on Meaning and Definite Descriptions" (*Ratio,* June, 1975) and "On Explicating Existence in Terms of Quantification" (*Logic and Ontology,* ed. M. K. Munitz, N. Y. U. Press, 1973) are earlier and in some cases flawed versions of portions of the present chapters 3 and 5.

I should like to thank Joseph Margolis, and Michael Lockwood, and William Tirrill for their help in making this monograph possible, and my students at Queens College and at the Graduate Center for their many stimulating questions and criticisms. Special thanks to Ann Sands for her help with the typing. Last of all I must express my gratitude to the Research Foundation of CUNY for providing me with a PSC - BHE research grant and to Dean Albert M. Levenson for his help and encouragement.

Alex Orenstein,
Queens College,
CUNY.

Contents

Contents

Introduction

There is a widespread view in philosophy that existence claims should be understood in terms of the logic of quantification. It is the purpose of this monograph to examine, criticize, and present alternatives to this view. I will try to show that linking existence and quantification has undesirable consequences for both logic and metaphysics. I begin by tracing the history of this treatment of existence from its inception in the work of Frege to its culmination in the writings of Quine. While doing this I note various claims about existence construed in this way, e.g., that 'exists' is not a predicate.

In Chapter Two, I shall demonstrate that there are ways of reading the quantifier so that it has no connection with existence. One of these non-existential readings, the substitutional reading, has in fact played an important—although until recently largely ignored—role in the history of contemporary philosophy.

A second non-existential reading to be examined is a neutral, or predicative, 'is' reading; 'is' here does not entail 'exists' and it can be construed along meinongian lines. In Chapter Three, I formulate certain requirements for a logic of quantification, viz., that it be free of any existence assumptions, and show how these requirements can be met by each of the different readings. Semantic considerations provide the grounds for these different readings and so in Chapter Three I will also consider the semantic frameworks that are part of the justification of these readings.

Chapter Four is a critical examination of the claims made about the notion of existence by proponents of the existential reading. Many of them have claimed that they were following Kant, who said that 'exists' is not a real predicate. Indeed this supposed link with Kant has been used to provide a historical precedent for the existential reading. I shall critically examine Kant's doctrine and find that it has little to do with the quantificational view of existence. Furthermore, there seems to be no clear precedent before Brentano or Frege for connecting quantification and existence. In Chapter Four and Five, I criticize the various claims about existence based on the quantificational account. These claims include the following: existence is not a meaningful predicate, it is a tautological predicate, it is a higher order property, etc. In the end I find no compelling reason that existence should not be viewed as an especially interesting non-logical notion, i.e., as a bonafide predicate constant. In Chapter Five, I examine the work of Quine, since he is the most serious defender of the position that quantification can serve as an explication of existence and indicate how his explication is inadequate. Throughout, I argue for the advantages of the substitutional treatment of quantification.

Existence
and the
Particular Quantifier

Chapter One

Quantification and Existence

"Existence is what existential quantification expresses."[1]

"Quantification is an ontic idiom par excellence."[2]

W. V. Quine

This chapter is devoted to the history of quantification in the nineteenth and twentieth centuries; of special interest is how the particular quantifier came to be linked with the notion of existence. In tracing the development of this existential reading of the quantifier, I shall adopt the attitudes of its proponents.

A. The Development of a Language of Quantifiers

'All' and 'some' and their equivalents have been central topics of logic ever since its inception. In Aristotle's syllogistic, their logical properties are treated in connection with the four types of categorical sentences, the familiar A, E, I, and O forms. Except for some instances in the Middle Ages, it was not until the first half of the nineteenth century that these expressions were systematically studied in other contexts, beginning with discussions about the quantification of the predicate. In place of the A form 'All A's are B's', we now have two forms: 'All A's are all the B's' and 'All A's are some of the B's'. Another relevant extension of logic in the nineteenth-century was in the area of relational sentences.

Yet a further factor to be taken into account was the increasingly more systematic use of quantifiers in mathematics. An important example of this is found in the idea of a limit, which had been introduced by Newton and Leibniz but was not clearly characterized until about the time of Cauchy. The definition of a limit involves the use of iterated quantifiers i.e. these would be symbolized today as '(x) $(\exists y)$ (z) $(...)$'. In his *Cours d'Analyse* (1821) Cauchy strove to duplicate in this area the rigor till then associated with Euclidean geometry. In some of his proofs Cauchy used natural language expressions such as 'for every x', 'there exists an x', 'one can find an x' to make explicit and to prove assumptions which had hitherto been regarded as self-evident. Eventually the definition of a limit attributed to Cauchy was given its present day formulation by Weierstrass in his lectures about 1860. By this time the use of these natural language quantifiers had gained greater currency, and they were employed more systematically. Even earlier than Cauchy, Bolzano in a work dating from 1804, *Betrachtung über einige Gegenstande der Elementar-geometrie*, and then more so in 1817 in *Rein analytischer Beweis . . .* , adopted standardized natural language quantifiers to state mathematical theorems and to construct proofs. However, Bolzano was regarded more as a philosopher and so appears to have had less influence on mathematical research in the nineteenth century than Cauchy. The work of Bolanzo, Cauchy and Weierstrass in the history of the logic of quantification parallels a pedagogical technique found in some textbooks of the new logic. Such texts require students to translate ordinary language statements into a standardized natural language quantificational format and only then into symbolic notation. In the nineteenth century the use of standardized natural language quantifiers provided a basis for developing a symbolic notation which would make matters of quantification even more perspicuous.[3]

These elements set the stage for the study of quantifiers in their full generality. Frege and Peirce working independently of each other were heirs to this common tradition, and brought this exploration to its culmination. It is to the credit of both of them to have developed a truly comprehensive symbolism for logic, a consequence of which is the treatment of 'all' and 'some' in this comprehensive context. Newtonian mechanics is a standard example of a great step forward in the history of science because it unified the hitherto disparate subjects of celestial and terrestrial mechanics. The new logic accomplished the same sort of synthesis. It provided a single notation suitable for all the forms mentioned above. The hitherto disparate logic of terms (i.e. the syllogistic) and the logic of propositions were now encompassed in a single system that was also a suitable vehicle for expressing mathematical concepts.

Both Peirce and Frege were well aware of the importance of this new notation. Peirce, in a piece significantly entitled "On the Algebra of Logic, A Contribution to the Philosophy of Notation" (1885), says:

> In this paper I purpose to develop an algebra adequate to the treatment of all problems of deductive logic, showing as I proceed what kinds of signs have necessarily to be employed at each stage of the development. I shall thus attain three objects. The first is the extension of the power of logical algebra over the whole of its proper realm. The second is the illustration of principles which underlie all algebraic notation. The third is the enumeration of the essentially different kinds of necessary inference; for when the notation which suffices for exhibiting one inference is found inadequate for explaining another, it is clear that the latter involves an inferential element not present in the former. Accordingly the procedure contemplated should result in a list of categories of reasoning, the interest of which is not dependent upon the algebraic way of considering the subject. I shall not be able to perfect the algebra sufficiently to give facile methods of reaching logical conclusions: I can only give a method by which any legitimate conclusion can be reached and any fallacious one be avoided. But I cannot doubt others, if they take up the subject, will succeed in giving the notation a form in which it will be highly useful in mathematical work. I even hope that what I have done may prove a first step toward the resolution of one of the main problems of logic, that of producing a method for the discovery of methods in mathematics.[4]

Frege was also aware of the importance of the new notation, as reflected in the title of his work *Begriffschrift, a Formula Language, Modeled upon That of Arithmetic for Pure Thought* (1879). In the preface he singles out Leibniz's quest for a *lingua characterica* as a precursor of his own search for an adequate symbolism. Like Peirce, Frege asserts that a universal notation must be adequate to the needs of logic.

> Its first purpose, therefore, is to provide us with the most reliable test of the validity of a chain of inferences and to point out every presupposition that tries to sneak in unnoticed, so that its origin can be investigated. That is why I decided to forgo expressing anything that is without significance for the *inferential sequence*. In section 3 I called what alone mattered to me the *conceptual content* [begrifflichen Inhalt]. Hence this definition must always be kept in mind if one wishes to gain a proper understanding of what my formula language is. That too is what led me to the name "Begriffschrift."

Then in section 3 we find

> Now, all those peculiarities of ordinary language that result only from the interaction of speaker and listener—as when, for example, the speaker takes the expectations of the listener into account and seeks to put them on the right track even before the complete sentence is enunciated—have nothing that answers to them in my formula language, since in a judgment I consider only that which influences its *possible consequences*. Everything necessary for a correct inference is expressed in full, but what is not necessary is generally not indicated; *nothing is left to guesswork*.[5]

Frege compares this "ideography" to ordinary usage by likening it to the accuracy of a microscope as compared with the ordinary eye. For everyday uses the eye serves well enough but to achieve scientific goals the microscope is indispensable. Like the microscope, Frege's *Begriffschrift* is a special instrument of use in special areas. Like Leibniz, Frege wanted a universal language, but unlike his predecessor did not expect to reach it in a single step. Instead he hoped to arrive at an approximation by first finding a comprehensive language for scientific purposes and then trying to extend it. Jean van Heijenoort has singled out Frege's claim of comprehensiveness for this new notation as a very important aspect of Frege's conception of logic: its universality.[6]

What is this new notation and exactly how does it differ from the older ones? We shall concentrate on the following two points—the treatments of relations and the clearer conception of the quantifiers. According to the older tradition, a sentence such as 'John is taller than Mary' was divided into a subject and predicate, 'John' being the subject and 'is taller than Mary' or 'taller than Mary' or some such variant as the predicate. Frege disparages this subject-predicate distinction and proposes instead one of functor and argument. To some extent it seems to be merely a matter of terminology whether we construe this as abandoning or as merely revising the subject-predicate distinction.

According to Frege, 'is taller than' is the function word or predicate (later he referred to it as a concept word) and 'John' and 'Mary' are the arguments or subjects.[7] Peirce, who sometimes uses the term 'indices' to mean 'arguments-subjects', makes the following remarks about this innovation.

> The algebra of Boole affords a language by which anything can be expressed which can be said without speaking of more than one individual at a time. It is true that it can assert that certain characters belong to a whole class, but only such characters as belong to each individual separately. The logic of relatives considers statements involving two and more individuals at once. Indices are here required. Taking, first, a degenerate form of relation, we may write $XiYj$ [i.e. more familiarly Xi & Yj] to signify that X is true of the individual i while Y is true of the individual j. If Z be a relative character Zij will signify that i is in that relation to j. In this way we can express relations of considerable complexity.[8]

The second point to note is the clearer conception of 'all' and 'some'. Within Boolean algebra, the words 'all' and 'some' function as operations on classes. Universal sentences are interpreted in terms of class inclusion, while particular sentences are understood as cases of class intersection. The quantifiers' function is incorporated into that of the copula which now assumes the role of a sign for class inclusion. Furthermore the subject and predicate expressions must designate classes. With Frege and Peirce 'all' and 'some' can be applied to individuals as well. This is in effect the recognition that one of the most distinctive functions of the quantifiers is that they are operators with respect to individual variables.

The history of logic reveals that a similar reading of quantifiers existed in some medieval theories of supposition.[9] This suggests an instructive partial analogy with the history of the propositional calculus. Just as Frege is to be credited not with discovering a new branch of logic, the propositional calculus, but rather with independently working out what is found in Stoic logic and the medieval theory of consequences, so the new conception of the quantifiers is like one aspect of the theory of supposition. But there is a crucial difference between the modern and the earlier theories of propositional and quantificational logic. Whereas the Stoic and medieval work in propositional logic yielded a relatively adequate symbolism, there is nothing in the theory of supposition quite like the new quantificational notation. In his history of logic, Bochenski records one of the most important consequences of this. "In contrast to the Aristotelian tradition, these quantifiers are conceived as separate from the quantified function and its copula, and are so symbolized." And then after comparing Albert of Saxony with Peirce we find: "Quite new, on the other hand, is the clear separation of the quantifier from the formula quantified."[10] This might be rephrased as: 'A notation which clearly separates the quantifier from the quantified formula is quite new'. The scholastics were well aware of the fact that 'all' and 'some' were syncategorematic expressions. Only in the new notation, though, was their distinctive function of binding variables recognized explicitly and exhibited in the symbolism itself.

As Peirce puts it

In order to render the notation as iconical as possible we may use Σ for *some*, suggesting a sum, and Π for *all*, suggesting a product. Thus $\Sigma i X i$ means that X is true of some one of the individuals denoted by i or

$\Sigma i X i = X i + X j + X k + \dots$ ["+" should be read as "or"].

In the same way $i X i$ means that X is true of all these individuals, or $\Pi X i = X i X j X k \dots$[11]

On this point Peirce is following directly in the footsteps of nineteenth century mathematicians. For instance, the sign he uses for sum—'Σ'—is to be found in the definition of an integral as an infinite sum of all values of a function. While it was Mitchell, Peirce's student, who was first to treat universal and particular propositions as products and sums, it was Peirce who in the above passage actually adopted the mathematical signs 'Π' and 'Σ' as quantifiers. Peirce and Mitchell were among the first to realize that 'all' and 'some' are the same sort of variable binding operations as are found in mathematics.[12] Soon thereafter Frege intro-

duced a special sign to render universal quantification and later treated
the notion of 'some' derivatively. When we couple this new clear conception of the role of the quantifiers with a symbolism for relational
sentences, then the study of 'all' and 'some' can be explored in their
full generality.

The orders of discovery and of the popularization of these concepts
differ. Although Frege (1879) is to be credited as the first to introduce
the quantifiers in their modern form (as well as to provide an axiomatization for them), his work was not generally assimilated until after that
of Peirce (1883) and Peano (1889). Peirce's notation for the quantifiers (ΠxFx) & (Σx) was taken up by Schröder and adopted by
Lukasiewicz and some other Polish logicians and philosophers. Peano's
notation was incorporated into Whitehead and Russell's *Principia
Mathematica*. Whitehead and Russell use '(x)', '$(\exists x)$' as their symbols
and it is primarily through the *Principia* that the new notation came into widespread use.

B. Quantification and Existence in Frege and Russell

One of the foremost philosophical uses of these new notions is their
application to questions about the nature of existence. In the
remainder of this chapter I will trace the development of what has
come to be the most accepted view of existence and will describe some
of the forces that have led to this widespread acceptance. It is somewhat ironic that, while most of Frege's distinctive contributions to
philosophical logic have only recently come to be appreciated, his
thoughts on the quantifiers and existence came via Russell directly into the mainstream of analytical philosophy. Let us turn then to Frege's
characterization of the quantifiers in the *Begriffschrift*. There he introduces (in the section entitled 'Generality') what has come to be called
the universal quantifier and then derives from that an account of existential quantification. '⊢' has already been introduced as a sign for
assertion, i.e. as signifying that that which follows it is what is being asserted.

> In the expression of a judgement we can always regard the combina-
> tion of signs to the right of ⊢ as a function of one of the signs occur-
> ing in it. *If we replace this argument by a German letter and if in*

the content stroke we introduce a concavity with this German letter in it, as in

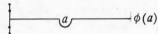

this stands for the judgement that, whatever we may take for its argument, the function is a fact.

He continues:

The horizontal stroke to the left of the concavity in

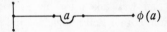

is the content stroke for the circumstance that, whatever we may put in place of a, $\phi(a)$ holds; the horizontal stroke to the right of the concavity is the content stroke of $\phi(a)$, and here we must imagine that something definite has been substituted for a.[13]

Frege is here noting the connection between quantification and substitution, a subject which will occur again in Russell and to which we shall devote the major part of Chapter Two. After exploring possible combinations of the symbolism for the quantifier and conditionals, Frege turns to combinations with negation (signified by '|').

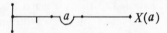

means

that we could find some object, say Δ, such that $X(\Delta)$ would be denied. We can therefore translate it as "There are some objects that do not have property X."

The meaning of

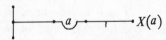

differs from this. This formula means "Whatever a may be, $X(a)$ must always be denied", or "There does not exist anything having property X", or if we call something that has property X an X, "There is no X".

We can translate the last formula as "There are Λ."[14]   From the outset, Frege forges a link between the quantifier and existence.

In the *Foundations of Arithmetic* (1884), Frege again takes up existence. While investigating the nature of number, and of the number one in particular, he points out that oneness is a property of a concept rather than of a thing (later he will call a thing an object). Roughly speaking, concepts are determined by functions, e.g., for 'John is tall', 'is tall' determines a concept. 'John' on the other hand names the thing-object John, which falls under the determined concept. Frege then distinguishes first and second level concepts. The former apply to things while the latter apply to concepts (presumably first level concepts). In the next paragraph (53) an analogous second level concept is discussed. Starting with the negative existential sentence 'There exists no rectangular triangle', Frege points out that this "does state a property of the concept 'rectangular . . . triangle'; it assigns to it the number nought." Frege immediately goes on to discuss existence proper, stating that "In this respect existence is analogous to number. Affirmation of existence is nothing but denial of the number nought."[15]

Frege then offers a critique, grounded in this account of existence, of Anselm's ontological argument.[16]   Existence cannot be a property of God because, as a second level property, it is not applicable to things (objects).   Then, by logical analogy, he refutes the claim that oneness is a first level property. (The point of the analogy is to suggest that such a claim would be an instance of the fallacy of composition. Russell later presents a similar refutation of the view that existence is a predicate of individuals.)

As an afterthought to his discussion of Anselm. Frege offers some qualifying remarks about existence and oneness. He tries to distinguish his own view of existence from a Kantian account. In opposition to what he takes to be the Kantian view, Frege says that existence and oneness (1) can under special circumstances be inferred from concepts, and (2) can be component parts of a concept. With regard to the first he says

oneness cannot be used in the definition of this concept (God) any

more than the solidity of a house, or its commodiousness or desir-
ability, can be used in building it along with the beams, bricks and
mortar. However, it would be wrong to conclude that it is in principle impossible ever to deduce from a concept, that is from its component characteristics, anything which is a property of the concept.
Under certain conditions this is possible, just as we can occasionally
infer the durability of a building from the type of stone used in
building it. It would therefore be going too far to assert that we can
never infer from the component characteristics of a concept to existence or to oneness.[17]

Second, Frege says it would be wrong to hold as Kant did that existence is not a real predicate in the sense that it is never a component of
a concept. Frege holds that existence can be a component of a concept, though not of a first level concept, but only of a higher level one.

It would also be wrong to deny that existence and oneness can ever
themselves be component characteristics of a concept. What is true
is only that they are not components of those particular concepts to
which language might tempt us to ascribe them. If, for example, we
collect under a single concept all concepts under which there falls
only one object, then oneness is a component characteristic of this
new concept. Under it would fall, for example, the concept "moon
of the Earth", though not the actual heavenly body called by this
name. In this way we can make one concept fall under another
higher or, so to say, second order concept.[18]

One other section of this work (29) is important for our purposes.
Here too Frege questions whether the number one is a property of objects. If 'one' named such a property, he claims, it would be a universal
property and then the predication of such a property would be senseless. Of course the same reasoning could be applied to 'exists', which
he equated with non-nullity (at-least-oneness). That everything is one
ensures that if oneness is a property of individuals, then it is a universal
property. Frege next invokes the law of inverse variation of extension
and intension. According to this law a predicate with a universal extension would have no intension. Such a word would not modify the
description of any subject.

In "Function and Concept" (1891) Frege changes his terminology, speaking of second *level* rather than second *order* concepts, and of first and second level functions. He explicitly informs us that

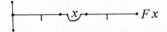

(in the more familiar notation '~(x) ~Fx') is called an 'existential sentence' and is a function of a function.[19] The essay "On Concept and Object" (1892) is concerned with distinguishing the two notions mentioned in the title. Salient to a history of the treatment of existence is the pronouncement that first level singular existence sentences are meaningless. This point is independent of Frege's earlier remarks about universal predicates; it is the outcome of a sharp distinction between concepts and objects, so that what is predicable of the one is not of the other.

> I do not want to say it is false to assert about an object what is asserted here about a concept; I want to say it is impossible, senseless, to do so. The sentence "There is Julius Caesar" is neither true nor false but senseless. . . .[20]

Also of some interest is the remark that the same thought (proposition) can be asserted in different ways and that in particular 'There is at least one square root of 4' could be put as well by 'The concept square root of 4 is realized'.

I have taken the trouble to go into Frege in this way because this portion of his work entered the mainstream of British and American analytical philosophy years before he was made available in the translations of Austin, Black, Geach, et al., and in the Frege studies of Church, Walker and Dummett. This link between Frege's treatment of existence and that of Russell, and through the latter with those of Moore, Kneale, Carnap, Ryle, Ayer, and others, seem to have gone unnoticed.

To trace the development of these ideas, I shall begin with Russell's views from 1903 to 1919. The relevant works are his *Principles of Mathematics* (1903), "On Denoting" (1905), "Mathematical Logic as Based on the Theory of Types" (1908), *Principia Mathematica* (1910), *The Philosophy of Logical Atomism* (1918), and the *Introduction to Mathematical Philosophy* (1919). Although Russell had read Frege when he wrote the *Principles of Mathematics* (indeed there is an appendix on Frege's logical and mathematical doctrines) still Russell had not yet assimilated Frege's thoughts on the quantifier and existence. It is in the famous "On Denoting" that we first find this material.

> My theory, briefly, is as follows. I take the notion of the *variable* as fundamental; I use "$C(x)$" to mean a proposition (more exactly a propositional function) in which x is a constituent, where x, the variable, is essentially and wholly undetermined. Then we can consider the two notions "$C(x)$ is always true" and "$C(x)$ is sometimes true."[21]

Interestingly enough, we do not find an explicit equation of 'sometimes true' and 'there exists' in this passage. The latter locution first appears when Russell distinguishes primary and secondary occurrences of definite descriptions, e.g., 'There is an entity which is now king of France and is not bald'. In "Mathematical Logic as Based on the Theory of Types," Russell introduces the quantifiers in the same way.

Late in this essay, after taking '$(x)\phi x$' as signifying "the truth of all values of a propositional function," '$\sim((x) \sim\phi x)$' is said to define "There is at least one value of x for which ϕx is true".[22]

In *Principia Mathematica* both quantifiers are taken as primitive and '$\exists x\phi x$' is explicitly introduced in terms of existence.

> The symbol "$(x) . \phi x$" may be read "ϕx always," or "ϕx is always true," or "ϕx is true for all possible values of x." The symbol "$(\exists x) . \phi x$" may be read "there exists an x for which ϕx is true," or "there exists an x satisfying ϕx."[23]

Though the phrase 'sometimes true' is still very much in use existence theorems are so called because of the presence of the quantifier. In addition a new symbol for class existence—'$\exists!$'—is introduced and defined in terms of quantification and class membership.[24]

$$\exists!a . = . (\exists x) . x \epsilon a$$

The same notation is extended in a systematically ambiguous way to relations, definite descriptions etc. (Actually Russell's use of '$\exists!$' is somewhat more complicated here; he uses this symbol only for certain special classes which in effect conform to the axiom of reducibility of his ramified theory of types. He calls these classes 'predicative' to distinguish them from the paradox-yielding impredicative ones. Later authors who subscribe solely to the simple theory of types use 'Ǝ!' in a less restricted fashion. (Unless indicated otherwise I shall use it in this less restrictive way.) In the chapter on descriptions in *Principia Mathematica* the existence sign 'E!' is defined for described individuals. Something described is distinguished from that which is immediately presented.

> It would seem that the word "existence" cannot be significantly applied to subjects immediately given; i.e., not only does our definition give no meaning to "E!x," but there is no reason, in philosophy, to suppose that a meaning of existence could be found which would be applicable to immediately given subjects.[25]

I shall put aside the epistemological distinction between the described and the immediately given. What interests me is that Russell too is making a claim that some simple singular terms are not suitable subjects for predications of existence. Indeed, to accept a sentence of the form '*a* exists' as significant, we must treat '*a*' as a disguised description.

It is in the *Lectures on the Philosophy of Logical Atomism* that Russell puts the greatest stress on the connection between the quantifier and existence. He begins to do this in Lecture V by pointing out that general propositions (particular propositions) and existential ones are the same. Kneale later makes the same point: "existential propositions are a subdivision of general propositions."[26] In Russell's Lectures, quantification is still discussed as a matter of propositional functions being always, sometimes, or never true. In Lecture V, Russell goes so far as to appear to equate these with the modalities. After describing the quantifiers in their always-, sometimes- and never-true way, he says:

One may call a propositional function

> *necessary*, when it is always true;
> *possible*, when it is sometimes true;
> *impossible*, when it is never true.[27]

A student who attended the lecture questioned this move, as did

Wittgenstein and Moore later.[28]   Russell in fact never seems to have made any use of this equation, but in his answer to the student he placed an interesting stress on the *root* notion:

> The first point I wish to clear up is this: I did not mean to say that when one says a thing exists, one means the same as when one says it is possible. What I meant was that the fundamental logical idea, the primitive idea, out of which both those are derived is the same. ... I used the word "possible" in perhaps a somewhat strange sense, because I wanted some word for a fundamental logical idea, for which no word exists in ordinary language, ... we say of a propositional function that it is possible, where there are cases in which it is true.[29]

The root idea for both 'exists' and this unusual notion of possibility is that of being sometimes true. So 'Lions exist' and 'Lions are possible' have in common that they are both predications about the propositional function 'x is a lion' to the effect that it is sometimes true.

Russell also provides an argument here that certain singular existence sentences (those whose subject is a proper name i.e., a simple singular term) are meaningless. The contention is that to apply existence to individuals is to commit the fallacy of composition-division. Frege had already exploited the idea for the notion of oneness. Consider the following arguments.[30]

A.   Men are numerous.
    Socrates is a man.
    Socrates is numerous.

B.   Men exist.
    Socrates is a man.
    Socrates exists.

C.   The things in the world exist.
    This is a thing in the world.
    This exists.

D.   This = the author of the homeric poems.
    The author of the homeric poems exists.
    This exists.

Though some of the above are of different logical form, they are all treated as instances of the same fallacy. In each case something is predicated properly in the premise of some whole (collectively) and then improperly in the conclusion of some part of the whole (distributively). In A, it is 'numerosity' of a class and then of a member of that class. In B and C, it is existence of a class or a propositional function and then of a member of a class or an individual satisfying the propositional function. In D, the error consists in going from the existence of a described individual to that of one introduced by a proper name.

There is a second, more influential argument that 'a exists' is meaningless. If 'a exists' were significant then it would have to be true (i.e. would be analytic). On the other hand if 'a exists' were false it would to be meaningless. For if 'a' were "really a name the question of existence could not arise, because a name has got to name something or it is not a name."[31] This argument makes use of the principle of significant negation, (i.e. to be significant, a sentence must have a significant negation) and relies on some version of a referential theory of meaning.

Finally, the *Introduction to Mathematical Philosophy* (1919) repeats many themes already considered. There are some important reflections on the ontological argument in the last chapter. (Actually some of this had already been touched on in "On Denoting," the *Principia*, and *The Lectures on Logical Atomism*. In all of these Russell points out that in 'God exists' and 'God is most perfect', 'God' must be taken as a definite description. In the *Principia* he demonstrates how, given his theory of descriptions, we can prove both the existence as well as the non-existence of such a being.) Here he says that if the ontological argument were correct, it would amount to a purely deductive proof of the existence of something.[32]

Russell then rejects Anselm's argument by invoking the fallacy of composition-division. It is said to be of the same logical form as argument D given above. But Russell now acknowledges that the same motives involved in rejecting the ontological argument—a purely deductive proof of an existence claim—should apply to logic as well. Hence there is something defective about some of the propositions of the *Principia*, e.g. '$x = x$' because they yield existence theorems, e.g. '$(\exists x)(x = x)$'. He then says that such propositions should not be given in that form, but should appear either as hypotheses or as consequences of hypotheses. The authentic theorems of logic are those whose truth is independent of the existence of any objects whatsoever.

Evaluating Russell's relationship to Frege, I find first that they each had two ways of construing the quantifiers in terms of existence and substitution. In some cases the two construals were explicitly equated. I have so far concentrated on their stress of the connection between

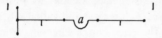

or '∃x' and existence. The other connection they make concerns the notion of substitution for propositional functions being sometimes or always true. Russell holds that existence is primarily a predicate of propositional functions, as Frege had held that it was a second level concept-function, i.e., that general existence sentences are of a higher level. Each claimed that singular existence sentences are meaningless. There is a difference, though. Frege treats definite descriptions (complex singular terms) as being of the same grammatical category as proper names (simple singular terms), and so presumably existence sentences with either of these as subjects would be senseless. For Russell the two belong to different categories and it is only the latter which yield nonsense.

C. Quantification and Existence From Wittgenstein to Quine

Though Wittgenstein's *Tractatus* (1918) is rich in insights, I shall only touch upon some of them here. There are signs that Wittgenstein saw a divergence between construing the quantifier '∃x' in terms of existence and in terms of substitution instances. These issues will be taken up in the next chapter. Here I shall note his claims that became part of the existential reading of quantification and for this reason I will also call attention to some of his followers' interpretations. Wittgenstein thought that his account of the quantifiers was an improvement on those of his predecessors. He thought, quite rightly, that what is common to '(x)Fx' and '(∃x)Fx' i.e., generality, is brought out by focusing on the nature of a variable.

5.522 What is peculiar to the generality sign is first, that it indicates a logical prototype, and secondly that it gives prominence to constants.

5.523 The generality-sign makes its appearance as an argument.[33]

The two quantifiers are seen as forming sums and products from amongst *all* the values of the variables. It is the idea of *all* the values or substitution instances of a variable which Wittgenstein seems to hint at. Today we would say that variables are associated either with a class of objects which constitute their range or with a class of substitution instances. This insight eventually was incorporated into formal semantics.

Better known is the claim that quantified sentences can be reduced to truth functional sentences, e.g. '(x)Fx' to a conjunction. Wittgenstein later admitted that this was a mistake.[34] Although there is actually very little in the *Tractatus* on existence, there is evidence that Wittgenstein was troubled about it. In his *Notebooks, 1914-1916*, we find qualms similar to those voiced by Russell.

> The question about the possibility of existence propositions does not come in the middle but at the first beginning of logic.

> All the problems that go with the Axiom of Infinity have already to be solved in the propositions "$(\exists x)x = x$".[35]

> 10.10.14.

Another relevant element in the *Tractatus* is his belief that words like 'object', 'thing', 'number' are pseudo-concepts. These words have the sole function of indicating what kind of variable we have (its range of values or substituends). Today this could be stated by saying that such pseudo-concepts function as devices for placing restrictions on quantifiers (indicating a relevant domain or substitution class) and not as ordinary predicates. Pseudo-concepts are redundant, in that there is no need for them in addition to the variable itself. Where these words are used as a sign for something over and beyond the variable, "nonsensical pseudo propositions result," e.g., 'There are objects' or '1 is a number'.[36] Carnap (1934) called these expressions universal words.[37] He generalized Wittgenstein's point, defining universal words or universal predicates as analytical predicates. They function either dependently as auxiliary symbols for a variable (e.g., 'There is an object x such that', or 'For all numbers x') "for the purpose of showing from which genus the substitution values are to be taken"; or independently as quasi-syntactical predicates in the material mode (e.g. 'The moon is a *thing*' is the material mode counterpart of ' "moon" is a thing word').[38] It is common to this tradition that such words are, if properly construed as placing restrictions on quantifiers, redundant and otherwise meaningless. One of Ramsey's (1927) defenses of Wittgenstein's view of '$(x)Fx$' as truth functional was that the sentence in which we try to express that besides Fa & Fb & Fc & . . . that a, b, c, etc. are *everything*, is nonsense.[39] In this context 'everything' or 'thing' is an improperly used universal word.

Ryle's influential "Systematically Misleading Expressions" (1931) furthers the view that 'exists' is a "bogus predicate."[40] Whereas Kant had put us on our guard about being misled by the form of one particular existence claim, viz. 'God exists', Ryle purports to expose all other existence claims for what they really are. In so doing he extends the list of universal words to include 'exists', 'is an existent' and others.

Earlier I pointed out that Russell argued that singular existence sentences were meaningless by appealing to the principle of significant negation. If 'a exists' is meaningful then '$\sim$(a exists), is also. But Russell reasoned that '$\sim$(a exists)' is not meaningful. Assuming that the meaning of 'a' is the object a, then—if the denial of existence were true—there would be no object a and both the word 'a' and the sentence in which it appears would be meaningless. Many philosophers of this generation held that assertions of singular existence are meaningless. The conclusion of Descartes's *Cogito*, viz. 'I am', is just such an assertion. In 1931 Carnap rejected this sentence as meaningless on Russellian principles.[41] Russell's argument, though, is not the only one that can be brought to bear here. One might formulate a principle of significant affirmation: if a sentence is meaningful, then so is its negation. Now 'a exists' is, if meaningful, analytically true (tautological) because the meaning of 'a' is the existent in question. It would follow, then, that '$\sim$(a exists)' is meaningful but a contradiction. In 1956, Ayer offered such an analysis of Descartes's conclusion.[42] Thus this is a new trend in which 'exists' in affirmative singular sentences is analytic, tautological, or redundant, and in such negative sentences is contradictory. This view had also been extended to general existence sentences. Variants in the same argument are stated in papers by Wisdom (1931), Ayer (1935), Moore (1936), and Broad (1939).[43]

Moore's paper, which appeared as a commentary on William Kneale's "Is Existence a Predicate?" (the latter is a lucid statement of the Russell-Frege view, but does not contain any new material), provided a wealth of details; I shall examine only two of his arguments for not considering existence a predicate. In the first, he concludes that 'exists' is meaningless as a predicate in general sentences.[44]

He might just as well have claimed that 'exists' in such cases is either tautological or contradictory, since the argument is really another instance of appealing to the principle of significant negation (or affirmation), but here applied to general sentences. According to the square of opposition, A- and O-form sentences are contradictories, i.e. negations, of each other. Hence, if an A-form sentence is significant, the corresponding O should also be significant. Moore compares what we will call "genuine" A- and O-form sentences about growling tame tigers with "pseudo" A and O sentences about existing tame tigers.

A   All tame tigers growl.

O   Some tame tigers don't growl.

Here the O is a significant negation of the A, hence the A is also significant. But now consider:

Pseudo A   All tame tigers exist.

Pseudo O   Some tame tigers don't exist.

Pseudo O′   There are tame tigers which don't exist.

Now Moore assumes in good Russellian fashion that 'some' or '$(\exists x)$' is interchangeable with 'There are', and thus that Pseudo O and Psuedo O′ are equivalent. But since Pseudo O′ is considered meaningless (or at least contradictory) then so is Pseudo O and therefore Pseudo A must be meaningless (or a tautology). Moore's arguments, as well as those of Wisdom, Ayer, and Broad, are designed to show that 'exists' is not a predicate (or not a "genuine" one). For if 'exists' were a predicate (like 'growls'), then it would be meaningful (or at least contingent). (A predicate might be said to be genuine if it were not definable in terms of purely logical constants, e.g. 'exists' appears to be definable in terms of the logical operation of quantification while 'growls' is not.)

In a second argument, Moore demonstrates that we can (or even *must*) dispense with using 'exists' as a predicate.[45] He takes Russell's reading of 'some', i.e. '$(\exists x)$', as 'sometimes true' quite literally and then sets up the following argument.

Upsala College
Library
East Orange, N. J. 070..

(1) Some tame tigers growl.

(2) For at least one value (substitution instance) 'x is a tame tiger and x growls' is true.

([2] is merely a substitutional reading of [1]. From [2] at least one sentence such as [3] follows)

(3) This is a tame tiger and this growls.

Now let us see what happens to (1').

(1') Some tame tigers exist.

(2') For at least one value 'x is a tame tiger and x exists' is true.

(3') This is a tame tiger and this exists.

Relying on the view generally accepted by 1936 that 'This exists' is meaningless (or redundant) and cannot serve as a conjunct, (3') is meaningless. 'This exists' must be dispensed with and similarly 'x exists' in (2') and 'exist' at the end of (1').

It is also important to note Moore's criticism of the view held by Russell that existence is some sort of higher level property of propositional functions. He calls our attention to the fact that while 'Some tame tigers exists' has the same meaning as ' "x is a tame tiger" is sometimes true', existence is not a property of propositional functions. For whatever we are saying of a propositional function when we say it exists, e.g., that 'x is a tame tiger' exists, we are not saying the same thing as saying it is sometimes true.[46]

Before concluding with the last major figure in the Russell-Frege tradition, I want to mention some other recent trends. We already find in Russell's *Lectures on Logical Atomism* the suggestion that one might express 'Something exists' by writing it as '$(\exists x)\, x = x$'.[47] In his 1914-1916 notebooks (12.11.14) Wittgenstein conjectured but rejected symbolizing the singular existence sentence 'a exists' as '$(\exists x)\, x = a$'. Some time later Heinrich Scholz expressed the *Cogito* as

$$\frac{(\exists x)\,(x = i\ \&\ Tx)}{(\exists x)\,(x = i)}$$

(i: I; T: thinks).[48] Reichenbach proposed in 1947 that 'x exists' be defined as '$(\exists y)\,(y = x)$', and said that this predicative use of 'exists' is not meaningless, but merely tautologous, presumably because $(\exists y)$ $(y = x)$ follows directly from the logical truth, or tautology $x = x$.[49] Salmon and Nakhnikian (1957) addressed themselves to some specific shortcomings in the account of Wisdom, Broad, and Ayer.[50]   These reflections led Salmon & Nakhnikian to view 'exists' as a significant predicate, which is at once universal, necessarily universal, and redundant. They suggest defining 'exists' in terms of '$x = x$'. With the latter, one can prove from 'Everything is self-identical' that everything exists, and then, that the former is necessary. 'Exists' would be redundant in the sense that '$x = x$' is a theorem of logic and provable from the null set of premises.

I turn now to Quine, who has more seriously than anyone else exploited this link between quantification and existence. Though others have already explained, and indeed argued for treating, existence in terms of quantification, Quine explicitly makes the use of quantifiers serve as an explication for ontological commitment, i.e., one's views about what there is. In "A Logistical Approach to the Ontological Problem" (1939) he quite clearly voices the views which have come to be his hallmark. He begins by addressing ontological questions such as "Is there such an entity as roundness?" Granting that expressions can be meaningful without being names, as syncategorematic expressions can, the question can be taken as inquiring whether 'roundness' is a name or a syncategorematic expression. But this merely raises the further question as to how to distinguish names from syncategorematic phrases. Quine's solution is to link names with both variables and operations on variables, viz. quantification. The ability to quantify with respect to an expression at once evidences (a) namehood for the expression and (b) ontological commitment.

> It is to names, in this sense, that the words 'There is an entity such as' may truthfully be prefixed. Elliptically stated: We may be said to countenance such and such an entity if and only if we regard the range of our variable as including such an entity. To *be* is to be a value of a variable.[51]

One of several consequences of Quine's program is that the use of different kinds of variables evidences different kinds of entities: propositional variables the existence of propositions, individual variables the existence of individuals, class variables the existence of classes, property variables the existence of properties, and so on. Another consequence is that in wedding quantification to existence he has divorced it from the early 'sometimes true'/'always true' locution. In recent years he has argued that the latter is not *real* quantification, is at best *derivative*, and gives rise to a *deviant* logic. In Quine the tradition urging a connection between '($\exists x$)' and 'exists' reaches its fullest and most critical expression.

Taking stock of the various claims writers who treated existence in terms of '($\exists x$)' have made, we feel that some have said 'exists' is not a predicate at all, some that it is a higher level predicate and some that it is not a higher level one. Different claims were made about singular and/or general existence sentences being meaningless or analytic. Finally implicitly and sometimes explicitly different claims were encountered about whether and how existence sentences were inferable, e.g. the ontological argument and existence theorems in mathematics and logic.

Notes

1. W. V. Quine, *Ontological Relativity and Other Essays* (New York: Columbia University Press, 1969), p. 97.

2. W. V. Quine, "Existence," in *Physics, Logic and History*, ed. W. Yourgrau (New York: Plenum Press, 1970), p. 92.

3. A. Grzegorczyk, *An Outline of Mathematical Logic*, (Dordrecht: D. Reidel, 1974), pp. 567-574. Grzegorczyk argues that this mathematical tradition was a more important influence in the development of quantification theory than the work done by Boole, Hamilton, et al. on such subjects as the quantification of the predicate.

4. C. S. Peirce, *Collected Papers*, ed. C. Hartshorne and P. Weiss, Vol. III (Cambridge: Harvard University Press, 1960), pp. 213-214.

5. G. Frege, "Begriffschrift," trans. S. Bauer-Mengelberg in *From Frege to Gödel, A Source Book in Mathematical Logic 1879-1931*, ed. J. van Heijenoort (Cambridge: Harvard University Press, 1967), pp. 6, 12. (Hereinafter referred to as *From Frege to Gödel*.)

6. J. van Heijenoort, *From Frege to Gödel*, p. 2, and "Logic as Calculus and Logic as Language" in *Boston Studies in the Philosophy of Science*, III, ed. Marx W. Wartofsky and Robert S. Cohen, (New York: Humanities Press, 1968), pp. 440-446.

7. Frege of course intended that the sentence could also be regarded as having 'John' as its argument and 'is taller than Mary' as its function word or similarly 'Mary' as the argument and 'John is taller than' as the function word.

8. Peirce, *Collected Papers*, Vol. III, p. 226.

9. E. A. Moody, *Truth and Consequence in Mediaeval Logic* (Amsterdam: North-Holland Publishing Co., 1953), pp. 43-53; J. Buridan, *Sophisms on Meaning and Truth*, trans. and with introduction by T. Kermit Scott (New York: Appleton-Century-Crofts, 1966), pp. 30-42.

10. I. M. Bochenski, *A History of Formal Logic* (Notre Dame, Ind.: University of Notre Dame Press, 1961), pp. 347,349.

11. Peirce, *Collected Papers*, Vol. III, p. 228.

12. For comments on the similarities with variable binding operators in mathematics see A. Mostowski, *Logika Matematyczna* (Warsaw: Wroclaw, 1948), pp. 45-48. He also has remarks on interpreting the quantifier geometrically. For the latter see Kuratowski and Tarski, "Logical Operations and Projective Sets" in A. Tarski, *Logic, Semantics, Metamathematics: Papers from 1923 to 1938*, trans. J. H. Woodger (Oxford University Press, 1956).

13. van Heijenoort, *From Frege to Gödel*, p. 24.

14. Ibid., p. 27.

15. G. Frege,*The Foundations of Arithmetic*, trans. J. L. Austin (New York: Philosophical Library, 1950), pp. 64-65.

16. Ibid., p. 65

17. Ibid., p. 64

18. Frege, *Foundations of Arithmetic*, p. 65. There is also a third way in which Frege differs from Kant. Kant uses 'existence' and 'actuality' interchangeably (as we shall see later on in Chapter Four), while Frege distinguishes the two. Geach has pointed out that for Frege existence is a higher level property while actuality applies to individuals. P. Geach and R. H. Stoothoff, "Symposium: What Actually Exists," *Proceedings of the Aristotelian Society*, Suppl. 42 (1968). In all of our references to Frege we are discussing his treatment of existence and not of actuality.

19. *Translations From the Philosophical Writings of Gottlob Frege*, P. Geach and M. Black, eds., (Oxford: Blackwell, 1952), pp. 37-38.

20. Ibid., p. 50.

21. B. Russell, *Logic and Knowledge* (New York: Macmillian Co., 1956), p. 42.

22. Ibid., pp. 83-85.

23. A. N. Whitehead and B. Russell, *Principia Mathematica to *56* (Cambridge: Cambridge University Press, 1962), p. 15. Hereinafter referred to as *Principia*.

24. Ibid., p. 29.

25. Ibid., p. 175.

26. *Readings in Philosophical Analysis*, H. Feigl and W. Sellars, eds., (New York: Appleton-Century-Crofts, Inc., 1949), p. 29.

27. Russell, *Logic and Knowledge*, p. 231.

28. There is a long tradition according to which necessity is understood in terms of universality. For instance see Kotarbinski on necessity in Aristotle's syllogistic; T. Kotarbinski, *Lecons sur L'Histoire de la Logique*, trans. into French by A. Posner (Warsaw: PWN-Polish Scientific Publishers, 1965), p. 8.

29. Russell, *Logic and Knowledge*, pp. 254-255.

30. Ibid., p. 233.

31. Ibid., p. 243.

32. B. Russell, *Introduction to Mathematical Philosophy* (London: George Allen and Unwin, Ltd., 1919), pp. 203-204; see also Russell, *Logic and Knowledge*, p. 240.

33. L. Wittgenstein, *Tractatus Logico-Philosophicus*, trans. D. F. Pears and B. F. McGuiness (London: Routledge and Kegan Paul, 1961), p. 103.

34. G. E. Moore, "Wittgenstein's Lectures in 1930-33," *Philosophical Papers* (New York: Collier Books, 1962), pp. 291-292.

35. L. Wittgenstein, *Notebooks, 1914-1916*, ed. G. H. von Wright and G. E. M. Anscombe, trans. G. E. M. Anscombe (New York: Harper Torchbooks, 1961), p. 10e.

36. Wittgenstein, *Tractatus*, p. 57.

37. R. Carnap, *The Logical Syntax of Language*, trans. A. Smeaton (Paterson: Littlefield, Adams and Co., 1959), sections 76 and 77.

38. *The Logical Syntax of Language*, p. 295.

39. F. P. Ramsay, *The Foundations of Mathematics and Other Logical Essays*, ed. R. B. Braithwaite (Paterson, Littlefield Adams and Co., 1960), p. 154.

40. G. Ryle, "Systematically Misleading Expressions," in *Logic and Language*, ed. A. Flew (Garden City: Anchor Books, 1965), pp. 19-20.

41. R. Carnap, "The Elimination of Metaphysics Through the Logical Analysis of Language," in *Logical Positivism*, ed. A. J. Ayer (Glencoe: The Free Press, 1959), p. 74.

42. A. J. Ayer, *The Problem of Knowledge* (Baltimore: Penguin Books, 1956), pp. 44-52.

43. J. Wisdom, *Interpretation and Analysis* (London: 1931), p. 62; A. J. Ayer, *Language, Truth and Logic* (New York: Dover Publications, Inc., 1935), p. 43; G. E. Moore, *Philosophical Papers* (New York: Collier Books, 1936), pp. 114-125; C. D. Broad, "Arguments for the

Existence of God I," Bobbs-Merrill Reprint Series in Philosophy (Phil-34), Indianapolis, 1939, pp. 17-24.

44. G. E. Moore, *Philosophical Papers* (New York: Collier Books, 1959), pp. 115-119.

45. Ibid., pp. 119-121.

46. Moore recognized that saying a propositional function exists and saying it is sometimes true are two quite different things. Not only do they differ in meaning but they are not even equivalent. If '*Fx*' designates an open sentence then that open sentence might exist without being sometimes true.

47. Russell, *Logic and Knowledge*, p. 240.

48. J. Slupecki and L. Borkowski, *Elements of Mathematical Logic and Set Theory* (New York: Pergamon Press, 1967), p. 127.

49. H. Reichenbach, *Elements of Symbolic Logic* (New York: The Free Press, 1947), pp. 332-333.

50. W. C. Salmon and G. Nakhnikian, "Exists As A Predicate," *Philosophical Review*, 66 (October, 1957), pp. 535-542.

51. W. V. Quine, *The Ways of Paradox and Other Essays* (New York: Random House, 1966), p. 66.

Chapter Two

Quantification Without Existence

"Do not forget that $(\exists x)Fx$ does not mean: There is an x such that Fx, but there is a true proposition 'Fx'."[1]

L. Wittgenstein

I now turn to examine two non-existential readings of the quantifiers. The first, the substitutional, is my major concern; the second, a neutral reading, will be discussed only towards the close of the chapter.

A. The Historical Sources for the Substitutional Reading

Although the existential reading discussed in Chapter One is by now quite firmly entrenched, it is not the only possible one. Before examining less well known ways of reading the quantifier in which '$\exists x$' is not linked with existence, we must adopt a more neutral attitude toward the expression '$\exists x$'. To refer to it as existential will not only prejudice us, but may actually beg the question of whether '$\exists x$' has some privileged connection with existence. This apparently simple point is of some consequence, and is perhaps the major obstacle—rhetorical, to be sure—that prevents one from acknowledging non-existential readings. I thus will speak of the *particular* quantifier, there certainly being ample tradition to sanction the term.

Probably the first explicit non-existential reading is Russell's locution 'sometimes true' found in all of his writings from 1905 through 1910. A case could be made for it occurring in Frege as well, or at least that Frege's view of '$\exists x$' as a function of a function leads in that direction. Though the seeds for two different readings are present in Russell's work, he never seemed to have questioned their compatibility. Later in the chapter I shall explain ways in which the two readings can diverge as well as ways in which they converge.

In this connection some of Wittgenstein's remarks are quite striking. For example, the quotation at the head of this chapter certainly shows that he had doubts about equating the two readings. The quotations in Chapter One on the subject of generality in terms of substitution instances rather than values can be interpreted in a similar spirit. Many of Wittgenstein's statements about quantification in the *Tractatus* appear in the context of his mistaken views about identity; these errors seem more plausible as instances of quite pervasive confusions about quantification and substitution.

Unfortunately the Polish schools of logic and philosophy (and especially that of Lesniewski) have remained rather isolated until recently. Here the expression 'particular quantifier' (when it is introduced at all, as opposed to using unnamed negation variants of the universal quantifier, e.g., '$\sim(x) \sim Fx$') is quite popular. We also find the words 'large and small quantifiers' which at first sound strange, e.g., "two quantifiers are used large and small, or rather universal and particular."[2] There are at least two reasons for their care in naming the quantifiers. In the first place these Polish writers apparently hold a substitutional view. In a work that was an important influence on Polish philosophy, *Elements of the Theory of Knowledge, Formal Logic and Scientific Method* (1929), Tadeusz Kotarbinski tells us that a sentence of formal logic

> opens with what is called the *quantifier*, which is an abbreviated symbol of an entire sentence . . . stating that the substitutions for the formula which follows are true. Thus, the quantifier is in itself a sentence which states that certain other sentences, namely those obtainable through the substitution of constants for the variables in a given formula are true. . . . Here, for instance, . . . $\Pi p(p \to p)$. Read: "For every p, if p then p," which means the same as: "Every sentence, obtained by the substitution of any sentence for the variable 'p' in the formula '$p \to p$', is true.

and a page or so later

But ... we should have to use not the large, but the small quantifier, by writing "$\Sigma p(p)$" and interpreting the quantifier thus: "Some of those sentences which are obtainable through the substitution of any sentence for 'p' in the following formula, are true sentences," or more briefly: "For some $p : p$."[3]

It is interesting to note that Ajdukiewicz, in his review of this book, called attention to a mistake made by Kotarbinski and one that has often been repeated about substitutional quantification, viz. a confusion of object language quantifiers and meta-linguistic ones.[4]

The second reason the Polish school distinguished the quantifier and existence can be found in Lesniewski's explicit definition of 'exists' as a predicate. Since he does not do this in terms of the quantifiers (or not solely with the help of them), he and those who accept his definition have something at stake in distinguishing 'exists' and '$\exists x$.' Lesniewski's systems of logic were employed by most Polish philosophers of that period, e.g., Kotarbinski.

We find traces of the two readings in British, American, and continental philosophers who followed in Russell's footsteps; they did not find any incompatibility in holding both views. Ruth Barcan Marcus (1962) and Henry S. Leonard (1964) more than any others, are responsible for making us aware of how the two views can diverge.[5] Though perhaps less influential, Mates and Geach were also well aware of some distinctive features of substitutional quantification.[6] It is interesting to conjecture why so many authors had a blind spot to the difference. For one thing Russell and many others were not clear about the difference between a value of a variable and a substituend for a variable. For instance, the value of a lowest level object-language individual variable would be some non-linguistic object, such as a particular dog Fido. It (or he) would be included in the domain of objects which our variable has as its range. On the other hand, a substituend for the same variable would be some linguistic object, e.g., an individual constant such as 'Fido'. Unless indicated otherwise, I will use the terms "substituend" and "value" of a given variable as mutually exclusive. Another reason was the utility at the time of treating existence in terms of quantification. This purported to solve such problems about existence as the ontological argument and was generally assumed to be in keeping with the accepted Kantian treatment of these subjects. Another use of this connection was in programs such as Quine's which try to clarify ontological questions by framing them in a logical notation.

B. Some Different Non-existential Readings of the Quantifiers

There are two non-existential readings to be considered—the substitutional view mentioned above and what might be called the neutral, or predicative 'is', reading. Below are four readings which are grouped under three headings. This array is convenient for discussing some confusions about substitutional quantification. We find later that the three types of readings can be used to mark more profound differences in interpretation. Additional readings could probably be clustered with one of these four.

	(x)	$(\exists x)$

I *The Substitutional Reading*

(1) Always true   Sometimes true
     True in all cases   In some cases (instances)
       (instances)

II *The Meta-linguistic Reading*

(2) For every expression   There exists an expression

III *The Neutral Reading*

(3) For all x For some x

(4) For all x There is an x
                                      (where 'is' has no
                                        existential import)

Authors treating quantification substitutionally have tended to use (1), (2) and at times (3).

Quine has used the terms 'substitutional' and 'referential' (or 'objectual') quantification.[7]   By 'referential'/'objectual' he means reference to a referent, i.e. usually a non-linguistic object, and not reference to a referend, i.e. a linguistic entity. Though this terminology and the classification it represents are in widespread use, I shall argue in a later section of this chapter that Quine's classification is seriously misleading.

Adopting this terminology for the moment, however, we find that a good way to understand the difference between the substitutional and referential/objectual views would be to construct a paradigm contrasting them.   Consider the sentence 'All humans are mortal', which is frequently symbolized as '$(x) (Hx \supset Mx)$'.[8]   On the substitutional view, we are concerned with the individual variable and its substituends. Here the substituends would most likely be individual constants.   The truth of the universal conditional would be a matter of the truth of every one of its substitution instances.   Another way of putting the matter is to consider the expansion of this sentence, e.g. '$(Ha \supset Ma)$ & $(Hb \supset Mb)$ & $(Hc \supset Mc)$ & . . . .'   For the finite case, a universal conditional is true if and only if its expansion is also true.   On the referential view 'All humans are mortal' is true when all the objects ranged over by 'x' are such that if any is a human then it is mortal.   Here we are concerned with all the values of the variable, while in the substitutional view we are concerned with all the substituends.   In a domain containing only Alfred, Betty, and Charles these three objects would, as human and mortal, suffice for the truth of the referential reading.   By contrast, the names 'Alfred,' 'Betty', and 'Charles' substituted for 'x' in '$Hx \supset Mx$' and each time yielding a true sentence, would suffice for the truth of the substitutional reading.

C. A Discussion of Six Confusions About
Substitutional Quantification

A number of confusions have surrounded substitutional quantification.[9] I shall consider six of them.

I. Substitutional Quantification Consists of
Defining the Quantifiers in Terms of Their Expansions

This confused notion can give rise to the following seeming rejection of the substitutional view.

(5)   Substitutional quantification consists of defining the universal quantifier by means of a conjunction and the particular quantifier by a disjunction, e.g. $(x)Fx =$ df Fa & Fb & Fc & . . . .

(6)   Only in finite cases do the expansions have the same logical

force as the quantifiers. Assuming that in logic we must deal
with the infinite case as well, then ' ... ' appended to the
expansion above would have to signify infinitely long conjunctions and disjunctions.[10]

(7)   The syntactical rules for well formed formulae (at least in the
       languages we shall be considering) allow only finite conjunc-
       tions and disjunctions.

Therefore,

(8)   Substitutional quantification fails to define the quantifiers.

The premise (5) is quite simply false. Wittgenstein was the only one
of all the authors to be considered who offered to define both quantifiers. He attempted to do so in terms of conjunctions and disjunctions,
but later rejected the effort. These conjunctions and disjunctions
would have had to be infinitely long.[11] Hilbert offered a definition of
the quantifiers but it rested on yet another primitive notion, viz. his
epsilon sign, and as such did not constitute a significant reduction. In
combinatory logic, the quantifiers can also be defined, but no adherent
of the substitutional view has taken this route. Of course it has been
customary for logicians—Frege, for example—to take one of the quantifiers, rather than both of them, as primitive, and then go on to introduce the other one derivatively. Interestingly enough neither Frege nor
Lesniewski ever introduced a special sign for the particular quantifier.

Even the somewhat more precise statement, " '$(x)Fx$' is true when
every substitution instance is," is not offered as a definition. This is a
rough but useful way of indicating the truth conditions, i.e., the
intended semantics for '$(x)Fx$'. If we were to take this point seriously
a parallel argument could be constructed for the referential/objectual
use of '$(x)Fx$'. According to this we say it is true when 'F' applies to
every object in your domain; but this way of calling attention to the
referential/objectual interpretation is not a definition either.

How then do we take care of the point raised in our argument about
the denumerably infinite case? The answer is that substitutional quantification requires at least a denumerably infinite amount of substituends, i.e., a number of the same size as the natural numbers. But this
number of expressions should not be cause for concern because in this

respect the language of a substitutional theory doesn't differ from that of a referential theory.   It is not unusual to assume that there are denumerably many expressions of the kind required.

II. Substitutional Quantification is Just a Species of the Referential Type, viz. Meta-linguistic Referential Quantification

The second confusion is that substitutional quantification is really nothing but a certain species of referential quantification, viz. referential quantification carried out at the meta-linguistic level. To illustrate this consider these inferences.

$$\frac{(x)Dx}{Df} \qquad \frac{Df}{(\exists x)Dx}$$ where 'D' abbreviates 'is a dog' and 'f' abbreviates 'Fido'

According to this confusion we would read the first as

(9)   *For every substituend* x, 'Dx' *is true.*
(10) Therefore, Fido is a dog.

and the second as

(11) *Fido is a dog.*
(12) Therefore, there exists a substituend for 'x' such that 'Dx' is true.

Here (9) and (12) are meta-linguistic statements and 'x' is no longer an expression of the object-language, i.e., it is no longer an object-language variable, either in the sense of having non-linguistic objects as values or in the sense of having object-language substituends.   It is a meta-linguistic variable and we can infer from the presence of 'exists' in (12) that the meta-linguistic quantification is existential.   More specifically, 'x' is treated here as a syntactical variable while the context indicates that its values are object-language substituends.   If this were a correct account of substitutional quantification we could construct the following argument.

(13) Referential quantification has existential import, i.e., it commits us ontologically.
(14) Substitutional quantification is a meta-linguistic species of referential/objectual quantification.

Therefore

    (15) Substitutional quantification has existential import and commits us ontologically.

From this one would conclude that substitutional quantification commits us to the existence of expressions. The remedy for this is to deny premise (14) and to refrain from using the meta-linguistic reading, i.e., 'there exists an expression' and its variants for substitutional purposes. This widespread confusion and its rectification were already noted by Ajdukiewicz in his review of Kotarbinski's treatment of the quantifier mentioned at the beginning of this chapter.

III. On the Substitutional Reading
There are No Object-Language Generalizations

    I will consider the locutions 'always true' and 'sometimes true' to render the substitutional view. But with these readings a third source of confusion appears. Consider the sentences

    (16) 'x' is a dog' is sometimes true.

or a paraphrase of it

    (17) In some instances x is a dog.

Now 'true' is often said to be a meta-linguistic predicate, and some have gone on to argue that sentences such as the above are meta-linguistic. On this view the first sentence above (16) would be meta-linguistic. Variants of 'sometimes true', as in (17), are not *prima facie* meta-linguistic, but if they are taken as equivalent to the first then they fare no better. It is claimed, then, that on this reading there are no object-language generalizations; and this is intended as a *reductio ad abusrdum* of substitutional quantification.

    One way out would be to adopt a neutral reading, e.g. 'for some x' or 'there is'. Both of these locutions yield object-language sentences. But as a natural way of rendering the substitutional idiom, 'sometimes true' is worth defending. Moreover, parallel situations arise in connection with other logical constants. An examination of one of these provides the necessary motivation for retaining 'sometimes true' while avoiding the *reductio* argument. Consider the propositional calculus formula '∼p' or the hybrid sentence '∼(Today is Sunday)' or indeed any sentence formed by using negation. Now consider the following argument.

(18) All the ways of reading '∼', i.e. negation, in English either explicitly use the word 'true', e.g., 'It is not true that today is Sunday' or equivalent ones such as 'it is false that', 'it is not the case that,' 'it is not so that' etc.

(19) 'True' is essentially a meta-linguistic predicate.

Therefore

(20) All negations are on the meta-linguistic level.

But now (20) is false on the usual interpretation of 'meta-linguistic'. Negation, '∼', is a way of constructing a more complex formula from a simpler one. But both the resulting negation and the unnegated original are on the same level of language. Since premise (18) appears to be empirically correct, we would have to modify (19). The attitude we should adopt toward 'true' is that it need not automatically and mechanically signify a higher-level sentence. Context must determine just how 'true' is being used. The same sort of error occurs when one insists that the presence of 'true' in 'always true' and 'sometimes true' as readings of '(x)' and '$(\exists x)$' signify that on this reading the quantified sentences must be taken meta-linguistically.

The view that sentences like 'It is sometimes true that cows are brown' are meta-linguistic might possibly originate in a confusion about the entirely plausible equivalence claim: It is sometimes true that cows are brown ≡ the sentential function 'x is a cow and x is brown' is true in at least one instance. Granting that the second sentence is meta-linguistic, it does not follow from its equivalence to the first, that the first is meta-linguistic too. To see how absurd such a position is, imagine someone reasoning in this way from Tarski's convention T, to the conclusion that there are no object language sentences. Perhaps the above is merely another illustration of one of Quine's points about semantic ascent. The element of semantic ascent, while frequently a feature of discussions of logical constants, is not distinctive of them. The fact that one can read or explain the particular quantifier as above is not good evidence that it is a linguistic or semantic notion. In Chapter Four I shall return to this confusion about equivalence claims and note its relevance to the thesis that 'exists' is a higher level or semantic property.

A discussion of the various readings of the quantifiers is seriously deficient without considering the quantifiers as parts of formal systems and the semantical frameworks which can be provided for these systems. Surely the justification of any of the readings would have to be done relative to the possible axioms or rules of inference for the quantifiers and the semantical truth conditions for quantificational sentences. In the next chapter I will make explicit the sort of formal systems I expect the quantifiers to be part of. Also to be considered are the several ways of providing truth conditions for them. At that time we will be in a better position to see which semantical frameworks coincide with which readings of the quantifiers. In this chapter, while trying to clarify and make a case for substitutional quantification, we are forced to jump the gun and consider some confusions as to the semantics of the quantifiers. In particular the next three confusions about the substitutional view pertain to the semantical systems in terms of which we explain these quantifiers.

IV. The Explanation of the Semantics for Substitutional Quantification Involves the Use of Referential Quantification and Hence the Former Presupposes the Latter

The fourth conclusion is that in giving the truth conditions for substitutional quantification we are making use of referential quantification, and thus the substitutional approach presupposes the referential one. This is not the same sort of mistake as the second confusion. There the substitutional variety was said to be merely a variant of the other type. Here we are told that in the meta-linguistic framework involved in supplying a semantical interpretation for the substitutional reading we will have to make use of referential quantification at the meta-linguistic level.[12]   It should suffice to point out that there is no *a priori* reason that the meta-linguistic quantifiers cannot also be construed in a substitutional sense.

V. The Explanation of the Semantics for Substitutional Quantification Involves a Notion of Instantiation, Which Has the Same Import as That of Referential Quantification

The fifth confusion is really not one, but a series of claims. It in-

volves saying that the attempt to explain the truth conditions underlying the substitutional reading rests on the notion of instantiation and
that the latter can be faulted. According to this account '$(\exists x)Fx$' is
true if the predicate 'F' is instantiated (or the property or concept F is).

There are at least two ways in which 'instance' and 'instantiation' are
ambiguous, even after we have distinguished predicate, property and
concept instantiation. Confining ourselves to predicates, the first sort
of ambiguity depends on whether a predicate like 'is a cow' is instantiated by a sentence such as 'Bossie is a cow' or by the fact of Bossie
being a cow.

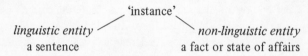

linguistic entity   *non-linguistic entity*
a sentence   a fact or state of affairs

The notion of substitution instance used to explain the substitutional
view is that of a sentence.

The second way in which 'instance' is ambiguous is in that either
'Bossie' or 'Bossie is a cow' may be taken as an instance.

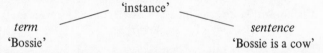

term   *sentence*
'Bossie'   'Bossie is a cow'

(Similar remarks apply to Bossie and the fact of Bossie being a cow). In
explaining the substitutional approach we would treat 'Bossie' as a substituend for the variable in the sentential function 'x is a cow' and 'substitution instance' would be restricted to 'Bossie is a cow'. The noun
'substitution' might generically cover both substituend and substitution
instance. So much for the sense of 'instance' involved in the substitutional reading.

Consider what might happen if we use 'instantiation' in an uncritical
fashion.

(21) S: '$(\exists x)\, x$ is a cow' construed substitutionally is true provided
   that 'is a cow' is instantiated.

(22) Bossie is the instance in question.

(23) Generally there has to be an object (existent) like Bossie
   serving as an instance in order for semantical sentences of the
   same sort as S to be true.

Therefore

(24) The substitutional reading is no more ontologically neutral than is the referential.

Here at the very least (22) is false, because the instance in question should be 'Bossie' and not Bossie.

VI. Substitutional and Referential Quantification Are Mutually Exclusive

Quine's introduction of the terms 'referential' and 'substitutional' quantification suggests that the two kinds are mutually exclusive. In the referential case, when we quantify, objects are referred to. As we have seen, this reference is the basis for connecting the particular quantifier with existence claims, and for ontological commitment being a matter of values of variables. The suggestion is that with substitutional quantification there are no questions of reference, but merely of substituends, constants, or, put slightly differently, merely questions of substitution instances. Going along with this suggestion is the view that substitutional quantification has nothing to do with questions of ontology and that the particular quantifier substitutionally construed cannot be used to express existence claims.

However, the distinction as made in the paragraph above is seriously misleading. Contrary to the above suggestion, substitutional quantifications can be referential, can have existential import, and can be a vehicle for ontological commitments. To illustrate this point, let us consider the sentence: All horses are white i.e., $(x)(Hx \supset Wx)$. On the substitutional construal a universal sentence is true iff every instance is true. Thus, 'If Silver is a horse than Silver is white', 'If Citation is a horse then Citation is white', and so on would all have to be true. Now there is no reason why the substituends (the singular terms) in these instances, 'Silver', 'Citation' etc. cannot be taken as referring to precisely the objects Silver, Citation etc. that are referred to on a non-substitutional reading. The quantification is substitutional because the quantifier is explained in terms of substituends and instances, but it is also referential because the substituends refer. In cases of this sort where the substitends refer to the same objects that are appealed to on a non-substitutional referential reading (as would be the case in explaining 'All horses are white' as true without appealing to substitution instances, e.g., as true when all the objects 'x' ranges over are such that if any are horses then they are white), then the two readings of the quantifiers coincide. As Kripke puts it in his recent paper on substitutional quantification:

> So, in the special case where (a) and (b) are postulated in the meta-theory*, there is indeed little difference between a substitutional quantifier and a referential quantifier ranging over the set of denotata.

> *(a) a totally defined denotation function is given for all the terms of L and (b) all formulae in L are transparent.[13]

In a similar spirit, 'Some horses are white', i.e. $(\exists x)(Hx \ \& \ Wx)$ can be construed substitutionally and referentially. The truth of the instance 'Silver is white and Silver is a horse' suffices for the truth of the substitutional generalization. If the singular term 'Silver' is used to refer and if our referents are limited to existents, then this substitutional reading and interpretation coincide with the existential reading and its referential interpretation. Thus substitutional quantification can be adapted to express existence claims and to record ontological commitments.

A proper understanding of the substitutional/non-substitutional distinction requires seeing that the distinction marks different kinds of truth conditions for general sentences, and that in the substitutional case, this distinction leaves open the question of how to formulate the truth conditions for the instances. When the truth conditions for the instances involve reference, e.g. when 'Silver is a horse' is true iff 'Silver' refers to one of the objects 'is a horse' applies to, the substitutional quantification inherits the referential force of its instances. (Substitutional quantification ceases to be referential when the truth of the instances is not a matter of reference. Examples of such cases will be provided later.) To reinforce the central point, that substitutional generalizations can inherit referential force from their instances, consider certain analogies between the quantifiers and truth functional connectives. Just as the generalizations

$(x)(Ax \supset Wx)$ and $(\exists x)(Hx \ \& \ Wx)$

can be understood in terms of their substitutional instances

$(Ha \supset Wa)$, $(Hb \supset Wb)$, etc.

and

$(Ha \& Wa)$, $(Hb \& Wb)$, etc.

so, the truth conditions for truth functional complex sentences such as the above conjunctions can be construed in terms of their constituents

Ha, Wa, Hb, Wb, etc.

The explanation of the truth conditions for the instances and those for the constituents can be independent of the exact formulation of the conditions for the generalizations and the conjunctions. A universal generalization is true if all its instances are, and nothing need be said here as to how the instances acquire their truth value. Similarly a conjunction is true when both conjuncts are, and nothing need be said as to how the conjuncts acquire their truth value. (Similar remarks can be made for conditionals, etc.) To maintain that substitutionally construed quantifications are never referential would be like saying that the truth of the claim that Socrates is human and Plato is human, as it is ordinarily understood, has no referential force because it suffices for the truth of the conjunction merely that both conjuncts be true. The conjunction, though, has referential force, i.e. those who make this claim refer to Socrates and Plato. The conjunction is a referential vehicle, not because of the truth conditions for the conjunction per se, but because of the reference that is part of the ordinary truth conditions for the conjuncts.

The convergence of substitutional quantification and reference, and of the existential reading of the quantifiers, has many precedents. To begin with, it provides a more generous way of understanding Russell's assimilation of the 'sometimes true' and the 'there exists' locutions. Somewhat more surprisingly, it furnishes a way of interpreting some of Quine's early remarks. In fact these remarks furnish a fine account of how the substitutional treatment of quantifiers and the referential treatment of instances can be put in the service of expressing existence claims by the use of the particular quantifier.

[I]nstead of describing names as expressions with respect to which existential generalization is valid, we might equivalently omit express mention of existential generalization and describe names simply as those constant expressions which replace variables and are replaced by variables according to the usual logical laws of quantification.

Here, then, are five ways of saying the same thing: "There is such a thing as appendicitis"; "The word 'appendicitis' designates"; "The word 'appendicitis' is a name"; "The word 'appendicitis' is a substituend for a variable"; "The disease appendicitis is a value of a variable." The universe of entities is the range of values of variables. To be is to be the value of a variable.[14]

I have distinguished substitutional from non-substitutional quantification, in that the former appeals to the notion of instances and the latter does not. But as we have seen there are also different variants of substitutional quantification. So far in this section, I have emphasized a variant which is referential. The convergence of substitutional quantification and "referential" is most functional when the substituends are straightforwardly referential. This occurs in the case of the non-vacuous names, etc., that serve as substituends for the variables of extensional first-order logic. Substitutional quantification and "referential" quantification diverge where the substituends are not clearly referential. Cases of this sort occur when the substituends are vacuous names, or names embedded in intensional contexts, or not names at all, e.g., predicate constants or sentences. Substitutional quantification for vacuous terms, intensional contexts and non-nominal positions can profitably be treated in a non-referential fashion. It is here that the substitutional reading of the particular quantifier has little if any connection with existence.[15]

I shall return to the question of clarifying the difference between substitutional, non-substitutional, and "referential" quantification in the next chapter. I shall keep Quine's terms 'substitutional' and 'referential' but shall not use them as mutually exclusive.

D. Substitutional Quantification and the Problem of Unnamable Objects

We now turn to a different sort of issue than those I have called "confusions": the question of whether there are in principle unnamable objects. If there were, '$(x)Fx$' could be substitutionally true but not substitutionally false. This problem takes its most serious form in the case of real numbers.[16]   We are at present proceeding on the assumption that the number of individual constants (and hence numerals) in our language is at most denumerable, i.e., can be placed in a one-to-one correspondence with the natural numbers. The amount of real numbers is non-denumerable, and so it seems that in this case we will have objects, viz. real numbers, which cannot all have names.

Abraham Robinson and Leon Henkin have offered a solution to this problem. Robinson proposes in *On the Metamathematics of Algebra* (1951) that we simply adopt a language with as many names as there are nominata.

> Let S be the set of objects in M which do not correspond to any object symbols under C, and conversely, let T be the set of object symbols in L which do not correspond to any objects under C. If the cardinal number of the set S is greater than the cardinal number of T, then we consider a language L′ which in addition to all the object symbols and relative symbols of L contains a set of object symbols T′ such that the cardinal number of T∪T′ is at least equal to the cardinal number of S (in accordance with what was said in the introduction we do not question the existence of such a language for any given structure).[17]

By 1963 he wrote of this innovation in the following way.

> Many of the topics included in "On the Metamathematics of Algebra"—such as the development of a non-countable language . . . are by now well-established and there is no need to justify their inclusion in the present book.[18]

Leon Henkin has also made use of such enlarged languages in his completeness proof for first-order logic.

> [T]he new method of proof which is the subject of this paper possesses two advantages. In the first place an important property of formal systems which is associated with completeness can now be generalized to systems containing a non-denumerable infinity of primitive symbols. While this is not of especial interest when formal systems are considered as logics—i.e., as means for analyzing the structure of languages—it leads to interesting applications in the field of abstract algebra.[19]

In yet another work, *Retracing Elementary Mathematics* (coauthored by Henkin), we find a notation presented for the real numbers and the outlines of a proof given that every real number is namable in such an enlarged language.[20] The central idea is to use infinite decimal expansions to name the real numbers. Such infinite decimals are frequently employed in high school mathematics to introduce the real numbers.

> We tend to think of a real number as an infinite decimal. For example we visualize $2^{1/2}$ as the infinite decimal 1.41421357..., where the three dots stand for infinitely many digits.[21]

To show that every real number is namable by an infinite decimal, Henkin proves that for every real number a there is an integer p and a real number k such that $0 \leqslant k < 1$, so that $a = p + k$. He also proves that for any a there is only one pair such as p, k. In other words every real number is in a one to one correspondence with the sum of a natural number p and a real number less than 1 but greater than 0 (e.g., $2^{1/2} = 1$ + some real number between 0 and 1). Since we can assume that we have a notation for natural numbers such as p our problem reduces to that of finding a notation for real numbers like k. Digits are defined by enumeration as the natural numbers 0, 1, 2, 3, 4, 5, 6, 7, 8, 9. A numeral of length n (where n is a positive integer) is defined as being a function, the domain of which is the natural numbers p for $p < n$ and whose value for each p is a digit. Intuitively speaking, numerals are treated in terms of rows or sequences such that every position in a given row is mapped onto a digit. For example the numeral '278' is of length 3 and is identified with the function f, consisting of the domain 0, 1, 2 such that $f(0) = 2$, $f(1) = 7$, and $f(2) = 8$. While an ordinary numeral is likened to finite row of digits an infinite decimal is likened to an infinite row of digits. Thus the notation '.333...' is associated with a constant function with N as its domain and whose every value is 3. Given this precise notion of an infinite decimal Henkin establishes that there is a one-to-one correspondence between infinite decimals and the real numbers, i.e., that every real number has an infinite decimal to name it and in fact exactly one. Though Henkin offers this proof as justification for the practice of introducing real numbers in high school mathematics by treating them in terms of infinite decimals, it clearly provides a suitable justification for substitutional quantification with respect to the real numbers.

Someone with constructivist scruples might object to the use of a non-denumerable language in the semantics of substitutional quantification. But this objection loses most of its force when we stop to consider that substitutional semantics resorts to this language only because the object language under consideration takes a non-constructivist attitude towards the real numbers. For instance, the language required by a substitutional semantics for an intuitionist's mathematics would be of a size that reflected a more constructive view of real numbers. In sum, there should be nothing strange about a non-denumerable amount of symbols in a language if one works with such amounts in one's mathematics; and if one doesn't work with such amounts in one's mathematics then one's language needn't contain them.

Other cases have been raised where substitutional quantification runs into problems in connection with set theory. Indeed Quine thinks that some are more serious than the problem about the real numbers.[22] I will not pursue these matters any further in this monograph, but will leave these issues to those more competent in set theory. However, even granting the possible inadequacy of substitutional quantification in these areas, I shall argue that there are other areas where objectual quantification runs into difficulties that can be avoided by the substitutional view. It may well be that it is mistaken to insist that only one interpretation of the quantifiers is correct.

E.   Some Differences Between a Substitutional and an Existential Reading

Having examined some confusions and a problem for the substitutional reading it is now time to explain a few ways in which it has been said to differ from the existential one. These differences are of three kinds.

I. The generalization rule.
II. Quantification and intensional contexts.
III. Quantifiers for variables of various grammatical categories.

The logic of quantifiers usually includes a rule for introducing the particular quantifier which is ordinarily called 'existential generalization'. Consider the use of the rule in the following well known argument.

(25) Pegasus is a flying horse.

Therefore

(26) $(\exists x)$ (x is a flying horse)   (25) Generalization

Grant that (25) is taken as true, as it is in some contemporary discussions (e.g. treat it as an instance that is true without assigning referents to its parts). If we read (26) as

(26a) There exists an individual x such that x is a flying horse.

then (26a) is literally false. This is a problem since as a rule of logic generalization should be truth preserving. On the other hand, if (26) is given a substitutional reading

(26b) It is sometimes true that there are instances of flying horses.

then (26b) is plausibly true by virtue of (25)'s constituting just such an instance. The two readings here appear quite incompatible.

Ruth Barcan Marcus has used this type of point to reply to several of the criticisms offered of quantified modal logic.[23] Quine has presented the following puzzle as a difficulty for modal logicians.

True (27) $\Box$ 9 is greater than 7.
True (28)   9 = the number of the planets.
False (29) $\Box$ The number of the planets is greater than 7.
    (30) $(\exists x) \Box$ (x is greater than 7).   from (27)–
                                                                    Generalization

It is given that (27) and (28) are true, and (29) is false. Reading (30) as

(30a)   There exists something necessarily greater than 7.

creates a puzzle. We have said that (30) follows from (27) by generalization. But (30) is problematic, for what object is necessarily greater than 7? To say it is the number 9, which is also the number of the planets, is incompatible with (29) which is false.

Marcus reads (30) as

(30b)   For some substituends, $\Box$ (x is greater than 7) is true.

On this reading (30) is true.

Another criticism of combining modalities and quantification has arisen in connection with the following formula (called the Barcan formula), found in some systems of modal logic.

(31) $\Diamond\,(\exists x)Fx \supset (\exists x)\,\Diamond Fx$

If this is read as

(31a)   If it is possible that there exists an x that has F, then there exists an x such that x possibly has F.

it appears to be false. The possibility cited in the antecedent does not preclude that there does not exist an object having that possibility. Consider the following instance:

If it is possible that there exist purple cows, then there exist possible purple cows.

The formula would sanction reasoning from the possible to what exists. This certainly is improper if we identify what exists with the actual or contingent, for we violate the principle that one cannot infer from the possible to the actual. Barcan Marcus, however, reads the formula as

(31b)   If it is possible that 'x is an F' is sometimes true for x, then it is sometimes true for x that it is possible that x is an F.

Here the doubts about the Barcan formula do not arise. In particular, the point that existence suggests actuality or contingency cannot be duplicated here, as the quantifier is not read existentially.

Another group of issues about quantifiers conveying existence claims arises in connection with modalities.

(32) $x = x$ Axiom
(33) $(\exists y)(x = y)$ (32)   Generalization
(34) $\Box\,(\exists y)(x = y)$ (33)   by the principle of modal logic that
                              if p is a theorem then so is $\Box p$.

Many contemporary authors read (33) as 'x exists'. But then by this reading (34) says that x necessarily exists.

(35) $(x)\,\Box\,(\exists y)(x = y)$ (34)   Universal Generalization

(35) is read as asserting that everything necessarily exists.

(36) $\Box (x)(\exists y)(x = y)$ (33)   Universal Generalization and if p
                                                 is a theorem then so is $\Box p$.

(36) tells us that it is necessary that everything exists. Each of these
sentences—(34), (35), and (36)—poses problems for those with scruples
about necessary existence. We have already seen some authors use arguments of this sort to prove that 'exists' is a universal predicate, or a
tautological or a redundant one. Look at the following inferences.

(37)   John is tall.
(38) $(\exists x)(x$ is tall$)$ (37)   Generalization
(39) $(\exists \phi)(\text{John } \phi)$ (37)   Generalization

It is as natural to go from (37) to (39) as it was from (37) to (38). But
if we insist that the quantifier be read existentially we must be prepared
to say what sort of objects 'ϕ' ranges over. The usual approach for this
case is to consider 'ϕ' as a set or property variable. Putting aside the
difficulties involved in giving an identity condition for properties, it
even seems strange to be involved with an ontology of sets at this point.
In a suitable universe with only three predicates (e.g., 'is tall', 'is dark'
and 'is handsome') where each is true of its object, (39) would be equivalent to

John is tall or John is dark or John is handsome.

Here no ontological oddities occur. On a substitutional reading '$(\exists \phi)$
(ϕj)', needn't have existential import and would be understood as saying
that 'ϕj' is true for some substituends of 'ϕx'.

In addition to predicate variables, we would sanction the use of
sentential variables, e.g.,

(40) $(p)(p \supset p)$

and—what strikes one as strange at first sight—yet other varieties of
quantification such as

(41) $(\exists f)(pfp)$.

(40) simply says that for all substitution instances, '$p \supset p$' is true and
(41) that for at least one substitution instance 'pfp' is true (or is sometimes true). It would take rather a forced reading to construe (41) existentially.

This possibility of quantifying with respect to diverse grammatical categories was critically and fully exploited by Lesniewski.[24] Particularly opposed to this tradition is Quine. For besides his concern for what exists he has attempted to limit quantification to variables whose only substituends if any would be singular terms. We shall return to this subject in the last chapter.

Earlier we departed from the historical approach adopted in Chapter One. The reason given was that many authors had not realized they were harboring two possibly divergent conceptions of quantification. This is no longer true. Writers such as Leonard, Barcan-Marcus, Geach, Leblanc, Dunn, Belnap, Gottlieb, Kripke and others are currently exploring the possibilities, problems and philosophical import of the substitutional conception.

In this section I have uncritically repeated some of the uses which proponents of substitutional quantification initially endorsed. In Chapter Three I will critically reexamine these claims. Thus an investigation of logics without existence assumptions will clarify the differences over the particular generalization rule, and a look at the different kinds of semantical framework will provide a better basis for judging the controversial Barcan formula. We will find that there are indeed differences of the three kinds mentioned but these needn't include every one of the claims initially made for substitutional quantification.

F.   A Neutral Reading

There is another non-existential reading of '$(\exists x)$' which need not be substitutional. The locution 'There is an x' in 'There is an x such that x is human' can be distinguished from 'There exists an x'. It is well known that 'is' has many uses—identity, class membership, predication, and of course existence. Granted that '$(\exists x)$' is construed as saying

(42)   There is an x (or there is a least one)

further argumentation is necessary to convince us that the 'is' here in (42) functions as the 'is' of existence and not that of predication. There are several ways of understanding the predicative 'is'. For our purposes, it suffices to note that some of these do not warrant inference from the predicative to the existential sense.

From: 'Every unicorn is a mythical creature' it follows neither that

Unicorns exist

nor that

Mythical creatures exist.

Similarly, from: 'Pegasus is a mythical creature' neither

Pegasus exists

nor

Mythical creatures exist

follows.

Many have noted this difference. There is, though, a tendency to go further than the evidence warrants. Thus, for example, John Woods claims that 'there is' does not commit us to existents but to beings.[25] It is one thing to revive the old distinction of being and existence. It is another to explain this distinction in Meinongian terms, taking as referents non-existent objects in addition to existent ones. Wood's claim is nonetheless instructive. Our own claim so far has been merely linguistic or logical, viz., that some functions of the predicative 'is' differ from 'exists'.

At least one of the points made about the divergence of the substitutional reading from the existential carries over to the 'there is' locution. (From here on I shall use 'there is' non-existentially unless quoting or explaining another author's views.) Consider the following inference.

True (43)   Pegasus is a flying horse.
     (44)   There is a flying horse. (Something is a flying horse.)

(44) would be considered true if we distinguish 'there is' and 'there exists'. The basis for holding that there is a flying horse or that something is a flying horse would be the truth of (43), that Pegasus is just such a horse.

Part of the purpose so far has been to present different readings for the quantifiers. I construe 'readings' as ordinary or natural language translations of the technical terms '(x)' and '$(\exists x)$', but the latter technical terms are part of the formalized theory.[26] We must see what is required of a formalized theory of quantification. What should its syntax be like? What sort of formulas should be theses (axioms or theorems) of the system? What alternative semantical approaches are suitable to these formalizations? Roughly speaking, semantics is the science of interpreting a formalized system, i.e., laying down truth conditions for its well-formed formulas. We will distinguish 'interpretations' from 'readings'. Readings occur in natural languages and not necessarily as part of a system. Interpretations occur in a system of semantics which can be formalized. Only after we have examined the syntax, system, and semantics for '(x)' and '$(\exists x)$', can we evaluate the several readings we have offered. Of particular importance here is seeing how the existential, substitutional, and neutral readings can be made to diverge. With this in hand we shall be in a better position to evaluate the claims made for the substitutional reading. We shall also be able to judge whether the neutral reading can be given a non-Meinongian justification.

Actually there is a certain amount of reciprocity between readings and interpretations of a formalism. Not only does the formalism allow us to judge the readings but the readings may furnish standards to apply to the formalism. Analogous cases abound. For example, consider 'true' or '⊃'. Ordinary readings of 'true' provide a material adequacy condition for Tarksi's formalized account of that notion. The formalization in turn suggests modifications in natural languages, e.g., that we cannot have expressions and their names on the same linguistic level. '⊃' is customarily read as 'If . . .then . . .', which provides certain constraints on a truth functional theory of '⊃'. But there still are disparities between '⊃' and 'If . . . then . . .' and these prompt us either to construct alternative accounts or modify our conception of 'If . . . then . . .' so that it approximates more to 'not both . . . and not . . .'.

Let us close this chapter by emphasizing the following results. There are, in addition to the existential reading of the quantifier, two seemingly legitimate non-existential readings, viz. a substitutional and neutral reading. We turn now to providing a theoretical justification for these different readings.

Notes

1. L. Wittgenstein, *Notebooks, 1914-1916*, ed. G. H. von Wright and G. E. M. Anscombe, trans. G. E. M. Anscombe (New York: Harper Torchbooks, 1961), 9.7.16, p. 75e.

2. T. Kotarbinski, *Elementy teorii poznania, logiki formalnej i*

methodologji nauk (Lwow: Wydawnictwo Zakladu Narodowego Imiena Ossolinskich, 1929), p. 160. This book has been translated by A. Wojtasiewicz and appears under the English title *Gnosiology: The Scientific Approach to the Theory of Knowledge* (New York: Pergamon Press, 1966), p. 134. (Hereinafter cited as *Gnosiology.*) Included in it are later essays by Kotarbinski as well as a 1930 review by K. Ajdukiewicz of the *Elements*.

Later references will be to the English translation except where indicated otherwise. The translator, Wojtasiewicz, has studiously avoided referring to the quantifiers as large and small. For instance compare his rendering of the material quoted "There are in use two quantifiers, universal and existential."

3. Ibid., pp. 132, 134-135.

4. Below are Ajdukiewicz's remarks on these passages in his review, *Gnosiology*, p. 530. The review is reprinted in the English translation.

> The opinion, formulated on pp. 129-130, concerning a sentence with a quantifier as a sentence about a sentence raises serious doubts in view of the fact that in such a case the sentential function within the scope of the quantifier would be taken *in suppositione materiali*. Further, it is wrong to say that "Thus, the quantifier is itself a sentence" (p. 132).   At the most it may be considered part of a sentence.

These observations and in particular the first one will be dealt with later in the chapter in connection with some confusions about the substitutional reading.

5. R. B. Marcus, "Interpreting Quantification," *Inquiry*, 5 (1962), 252-259; H. S. Leonard, "Essences, Attributes and Predicates," *Proceedings of the American Philosophical Association*, 37 (April-May, 1964), pp. 25-51.

6. B. Mates, "Synonymity," in *Semantics and the Philosophy of Language*, ed. L. Linsky (Urbana: University of Illinois, 1952), pp. 132-133; and P. Geach, "Quantification Theory and the Problem of Identifying Objects of Reference" in his *Logic Matters* (Berkeley: University of California Press, 1972), p. 144.

7. W. V. Quine, *Ontological Relativity and Other Essays* (New York: Columbia University Press, 1969), pp. 64 and 105. (Hereinafter cited as *Ontological Relativity.*)

8. There may be good reasons for not symbolizing this sentence in this way. For instance it is not plausible that the sentence 'All humans are mortal ' is about every individual x which is what '$(x)(Hx \supset Mx)$' is about.

9. For a discussion of some of these same confusions see M. Dunn and N. D. Belnap, Jr., "The Substitution Interpretation of the Quantifiers," *Nous*, 2 (1968), 177-185.

10. In this monograph I shall avoid some controversial issues in the philosophy of mathematics. In 6, I take a moderate stand by assuming that quantification must take account of at least as many objects as there are natural numbers, i.e., a denumerably infinite amount. Hence the above argument constitutes a problem for us. On other assumptions, such as strict philosophical finitism which holds that there are only a finite number of objects, the above argument would not apply.

11. There are at present some logicians who are working with infinitely long WFF's; see C. R. Karp, *Languages with Expressions of Infinite Length* (Amsterdam: North-Holland Publishing Co., 1964).

12. *Physics, Logic and History*, W. Yourgrau, ed. (New York: Plenum Press, 1970), p. 185. David Kaplan says the following while commenting on the views of Lejewski.

> I think the best way of understanding the substitution sense of the quantifier is as follows: We should here adopt Quine's device of semantic ascent and say that the sentence "$(\exists x)Fx$" should be understood in this way; not that "There is some individual x which runs" is true, but rather, there is some individual constant a such that the result of writing a before "runs" is true. I am using referential quantifiers in the metalanguage here of course.

13. S. Kripke, "Is There a Problem about Substitutional Quantification?" in *Truth and Meaning*, ed. G. Evans and J. McDowell, (Oxford: Clarendon Press, 1976), p. 351.

14. W. V. Quine, "Designation and Existence" in *Readings in Philosophical Analysis*, ed. H. Feigl and W. Sellars (New York: Appleton-Century-Crofts, 1949), pp. 49-50.

15. As a historical example of the use of non-referential substitutional quantification consider Ramsay's remarks on the difference between two types of propositional functions, those of first and of second order logic. I interpret Ramsay as trying to say that second order quantification involving propositional functions of propositional functions should be construed substitutionally but not referentially. Wherever he speaks of functions he means propositional functions.

> As so far there has been no difficulty, we shall attempt to treat functions of functions in exactly the same way as we have treated functions of individuals. Let us take, for simplicity, a function of one variable which is a function of individuals. This would be a sym-

bol of the form '$f(\hat{\phi x})$', which becomes a proposition on the substitution for '$\hat{\phi x}$' of any function of an individual. '$f(\hat{\phi x})$' then collects together a set of propositions, one for each function of an individual, of which we assert the logical sum and product by writing respectively '$(\exists\phi) . f(\hat{\phi x})$', '$(\phi) . f(\hat{\phi x})$'.

But this account suffers from an unfortunate vagueness as to the range of functions $\hat{\phi x}$ giving the values of $f(\hat{\phi x})$ of which we assert the logical sum or product. In this respect there is an important difference between functions of functions and functions of individuals which is worth examining closely. It appears clearly in the fact that the expressions 'function of functions' and 'function of individuals' are not strictly analogous; for, whereas functions are symbols, individuals are objects, so that to get an expression analogous to 'function of functions' we should have to say 'function of names of individuals'. On the other hand, there does not seem any simple way of altering 'function of functions' so as to make it analogous to 'function of individuals', and it is just this which causes the trouble.

16. For a discussion of alternate solutions to this problem see A. I. Fine, "Quantification Over the Real Numbers," *Philosophical Studies*, 19 (January-February, 1969), pp. 27-32.

17. A. Robinson, *On the Metamathematics of Algebra* (Amsterdam: North Holland Publishing Co., 1951), p. 20.

18. A. Robinson, *Model Theory* (Amsterdam: North Holland Publishing Co., 1963), p. v.

19. L. Henkin, "The Completeness for the First-Order Functional Calculus" in *The Philosophy of Mathematics*, ed. J. Hintikka (London: Oxford University Press, 1969), p. 42.

20. L. Henkin, W. N. Smith, V. J. Varineau and M. J. Walsh, *Retracing Elementary Mathematics* (New York: Macmillan Co., 1962), pp. 390-397.

21. A. H. Lightstone, *Symbolic Logic and the Real Number System* (New York: Harper and Row, 1965), p. 168.

22. W. V. Quine, *The Roots of Reference* (LaSalle: Open Court, 1973), p. 113.

23. *Contemporary Readings in Logical Theory*, I. M. Copi and J. A. Gould, eds. (New York: Macmillan Co., 1967), pp. 288, 289; R. B. Marcus, "Modal Logics: I, Modalities and Intensional Languages." *Boston Studies in the Philosophy of Science. Proceedings of the Colloquium of 1961/62.* Dordrecht (1963), pp. 97-104.

24. See T. Kotarbinski, *Lecons Sur L'Histoire de la Logique*, translated into French by A. Posner; (Warsaw: PWN-Polish Scientific,

1965), Chap. xxiii; especially sec. 4 entitled "Quantifiers Binding Variables From Other Semantic Categories," pp. 225-227.

25. J. Woods, "Essentialism, Self-Identity and Quantifying," in *Identity and Individuation*, ed. M. Munitz (New York: New York University Press, 1971).

26. Richard Martin has called for making the relation between expressions of a natural language and their symbolic correlates explicit in what he calls "rules of correlation". In the present case the problem is one of correlating natural language expressions such as 'some', 'there is', 'there exists' and 'sometimes true' on the one hand with the technical term '∃x' on the other. Martin in the following quotation (from *Logic, Language and Metaphysics* [New York: New York University Press, 1971], pp. 9, 96-99) talks of the English 'and' and the symbolic '·'. These remarks can be applied as well to 'some', 'there is' etc. and '∃x'.

What is needed now is an exhaustive classification of *all* uses of English words such as 'and' and 'or' and so on. *Rules of Correlation* would then be formulated as between contexts containing them and purely symbolic contexts containing the logical symbols. Nothing short of an exhaustive study of this kind could justify correlating the English 'and' in many contexts with the symbolic '·'.

Chapter Three

Quantificational Systems
and Their Semantics

Prior to or independent of the selection of a language-system L the quantifiers are without meaning. They are given meaning only with the formulation of L, i.e., by the explicit listing of the syntactical and semantical rules determinative of L.[1]

R. M. Martin

In this chapter I consider certain requirements for the rules of inference and the theses, i.e., axioms and theorems, of a theory of quantification, and try to provide semantical frameworks for the existential and non-existential readings of the quantifiers.

Part I   The Existential Reading

We turn to certain issues in the axiomatization or systematization of quantification theory. These shall be dealt with in three sections.

A. Quantification theory as a logic free of existence assumptions.

B. Arguments purporting to establish that we don't need a free logic.

C. Semantical frameworks for the existential reading.

A.   Quantification Theory as a
Logic Free of Existence Assumptions

It is customary to distinguish two senses in which logic should be free of existence assumptions:

I. The theses of logic should be true in all possible worlds, one of which is empty.
II. The principles of logic should apply to any constants, even vacuous ones.

A First Requirement for a Free Logic; Quantification and the Empty Domain

We have already noted that Russell realized it was a shortcoming of *Principia* that one could prove '$(\exists x)(x = x)$' as a theorem. A condition for understanding '$(\exists x)(x = x)$' existentially is that 'x' range over some domain. Furthermore, in order for '$(\exists x)(x = x)$' to be true the domain must contain at least one member. In this sense all such existentially construed sentences are true only in domains containing at least one member. Since then, the issue has been hedged in many texts by saying that a formula is logically true (or valid) if true in every non-empty domain. Motivated by the Leibnizian insight that the truths of logic should be true in all possible worlds, we ought to revise the above to true in every domain including the empty domain. Consider the following sentences which, on the older conception, counted as truths of logic.

(1) $(\exists x)(Fx \ v \sim Fx)$

(2) $(x)Fx \supset (\exists x)Fx$

Neither of these sentences is true for the empty domain, and they are but two of many. The principle according to which we decide on truth values with respect to the empty domain is to treat as true any well formed formula (WFF) whose left-most expression is a universal quantifier, the scope of which is the entire WFF, e.g. the antecedent of (2). The negation of such a WFF would be false; all WFFs with an existential-particular quantifier whose scope extends to the end of the WFF would be false since they are equivalent to the negated universal sentences. Sentence (1) is false for the empty domain, asserting as it does that there exists at least one individual such that $Fx \ v \sim Fx$. (2) is false because it is a conditional with a true antecedent and a false consequent.

We can solve the problem by modifying the axioms and/or rules of inference. I will consider two ways of doing so.

The first is derived from Quine's book *Mathematical Logic* and his article "Quantification and the Empty Domain."

In *Mathematical Logic*, Quine offers axioms for quantificational logic.[2] He uses Greek letters as meta-linguistic variables.

'a', 'β', 'γ'   range over individual variables

'ϕ', 'ψ', 'χ'   range over WFFs

'ξ', 'η', 'σ'   range over terms, e.g., individual constants

With these he presents the following system:

*100 If ϕ is tautologuous, $\vdash \phi$.   ('$\vdash \phi$' means that the closure of ϕ is
a theorem)

*101 $\ulcorner (a)(\phi \supset \psi) \supset . (a)\phi \supset (a)\psi \urcorner$.

*102 If a is not free in ϕ, $\vdash$ $\ulcorner \phi \supset (a)\phi \urcorner$.

*103 If ϕ' is like ϕ except for containing free occurrences of a'
wherever ϕ contains free occurrences of a, then $\vdash$ $\ulcorner (a)\phi \supset \phi' \urcorner$
or more simply $\ulcorner (a)\phi \supset \phi \beta/a \urcorner$.

*104 If $\ulcorner \phi \supset \psi \urcorner$ and ϕ are theorems, so is ψ.

With the above it would still be possible to prove $(x)Fx \supset (\exists x)Fx$. Consider the following.

(3) $(x)Fx$	Assumption for conditional proof
(4) $\sim(\exists x)Fx$	Assumption for an indirect proof
(5) $\sim\sim(x)\sim Fx$	4 Definition of $(\exists a)\phi$ as $\sim(a)\sim\phi$
(6) $(x)\sim Fx$	5 Double negation
(7) $(x)(Fx \supset f)$	6 Definition of $\phi \supset f$ as $\sim\phi$
(8) $(x)Fx \supset (x)f$	8 *101
(9) $(x)f$	3, 8 Modus ponens
(10) f	9 *103
(11) $\sim f$	Tautologous
(12) $(\exists x)Fx$	4-11 Indirect proof
(13) $(x)Fx \supset (\exists x)Fx$	3-13 Conditional proof

In "Quantification and the Empty Domain" Quine modified *103 (i.e., $\vdash \ulcorner(a)\phi \supset \phi\beta/a\urcorner$) by adding the condition that a must be free in ϕ.[3] With this restriction the move from line (9) to (10) would no longer be justified, since 'x' is not free in 'f', so the formula in question is not a theorem. Similar constraints would apply to attempts to prove '$(\exists x)(Fx \text{ v} \sim Fx)$'.

There is a different solution to this problem, which is much more widespread. It has been adopted, with variations, by several authors, e.g. Lejewski, Lambert, Hintikka.[4] I will state what these authors have in common and then give Hintikka's version. His modified rules of inference will serve as the basis for most of the discussion later in the chapter.

Consider the rules commonly known as universal instantiation and existential generalization. Using Quine's notation these can be stated as follows:[5]

The Traditional (Unmodified) Rules

Universal Instantiation

with respect to variables $\quad \dfrac{(a)\phi a}{\phi\beta} \quad$ where ϕa is like $\phi\beta$ except for containing a free wherever β occurred free in $\phi\beta$.

with respect to terms $\quad \dfrac{\phi\eta}{(\exists a)\phi a} \quad$ with a restriction similar to the one above.

Existential Generalization

with respect to variables $\quad \dfrac{\phi\beta}{(\exists a)\phi a} \quad$ where ϕa is like $\phi\beta$ except for containing a free wherever β occurred free in $\phi\beta$.

with respect to terms $\quad \dfrac{\phi\eta}{(\exists a)\phi a} \quad$ with a restriction similar to the one above.

Given these rules '$(x)Fx \supset (\exists x)Fx$' and '$(\exists x)(Fx \text{ v} \sim Fx)$' can easily be proved, the first by assuming the antecedent, instantiating and then generalizing, and the second either by generalization on the propositional calculus theorem '$Fa \text{ v} \sim Fa$' or by instantiation with respect to the theorem '$(x)(Fx \text{ v} \sim Fx)$' and then generalization. If we modify these rules in the following fashion none of the undesirable existence sentences can be proved as theorems.

Modified Rules

Universal Instantiation

with respect to variables $\dfrac{(a)\phi a}{\beta \text{ exists} \supset \phi\beta}$ (We shall tacitly understand in these and later rules the same sort of restrictions on substitution given above.)

with respect to terms $\dfrac{(a)\phi a}{\eta \text{ exists} \supset \phi\eta}$

Existential Generalization

with respect to variables $\dfrac{\phi\beta}{\beta \text{ exists} \supset (\exists a)\phi a}$

with respect to terms $\dfrac{\phi\eta}{\eta \text{ exists} \supset (\exists a)\phi a}$

This solution consists of supplying conditionals as the conclusion of the modified rules; the consequences of the conditionals were themselves the conclusions of the unmodified rules. The existence assumption is explicitly stated in the antecedent of the modified conclusion. This means that in order to prove an existence sentence, we have to have an existence sentence as a premise. The method is common to a number of "free logics,"[6] a fitting term coined by Lambert which I will use in a somewhat similar fashion. This solution can be viewed as a reworking of Russell's suggestion that "Propositions of this form [existential], when they occur in logic, will have to occur as hypotheses or consequences of hypotheses, not as complete asserted propositions."[7]

Variations of these modified rules can be obtained by defining or refining 'exists' in the antecedent 'a exists'. Some authors adopt Russell's sign '$\exists! a$', but use it in a different way than he did. Lejewski defines 'exists' in terms of one kind of Lesniewskian inclusion.[8] Some leave 'exists' as an undefined constant.[9] I adopt '$(\exists\beta)(\beta = a)$' as a working definition of 'a exists'. Indeed Hintikka proposed rules for a free logic which were formulated essentially in just this way.[10]

Hintikka's Modified Rules

Universal Instantiation

with respect to variables $\dfrac{(a)\phi a}{(\exists\gamma)(\gamma = \beta) \supset \phi\beta}$

with respect to terms $\dfrac{(a)\phi a}{(\exists\gamma)(\gamma = \eta) \supset \phi\eta}$

Existential Generalization

with respect to variables $\dfrac{\phi\beta}{(\exists\gamma)(\gamma = \beta) \supset (\exists a)\phi a}$

with respect to terms $\dfrac{\phi\eta}{(\exists\gamma)(\gamma = \eta) \supset (\exists a)\phi a}$

A Second Requirement for a Free Logic; Quantification with Vacuous Terms

The second requirement for a free logic is that there be no restrictions, save grammatical ones, as to the kinds of non-logical constants, i.e., descriptive ones, that the rules of logic apply to. To highlight the shortcomings of a logic not free in this sense, we could consider some of the rules of a "traditional" version of Aristotelian logic. According to the traditional square of opposition, an I-form proposition follows from an A-form proposition. But this rule is not truth-preserving when the subject term is vacuous, and the sentences are treated respectively as generalizations of conditionals and conjunctions, e.g.:

A   All unicorns are single-horned horses.

I   Some unicorns are single-horned horses.

We can show that the argument is not sound by taking the A-form sentence as saying truly that if anything is a unicorn then it is a single-horned horse, while the I-form makes the false existence claim that there exist unicorns and that they are single-horned horses. The solution given in most introductory texts points out that the rules apply only to referential (non-vacuous) terms, i.e., terms that refer to existing objects. This is surely incompatible with our belief that logic should be formal; if an expression is of the appropriate syntactical category, that alone should suffice for applying a rule of logic to it. For instance, the fact that an expression is a sentence is a syntactical point and should suffice to determine whether a rule of the sentential calculus can be applied to it without our having to know whether the sentence is true or false. The use of an individual constant is similar: it is not necessary that it have a designation for the constant to fall under an appropriate rule of logic.

Many authors thus reject much of "traditional" logic as containing spurious principles. But does modern logic with its unmodified rules fare any better? Consider what happens when one uses a vacuous singular term like 'Pegasus' in the unmodified rule of universal instantiation.

(14)   Everything is in space and time.
(15)   Pegasus is in space and time.

If we adopt a materialist point of view, (14) is true while (15) is evidently false. Or consider:

(16)   Pegasus is a flying horse.
(17)   There exists an x such that x is a flying horse.

For those who consider (16) true and (17) false, (16) is considered true on the grounds that it is taken as true in mythology, but (17) is false because it is assumed that Pegasus never has existed and never will. To say, as most texts do, that these rules do not apply unless we already know that the individual constants involved have a referent is to make the same mistake as did the "traditional" Aristotelians. By the same reasoning we should acknowledge that these unmodified rules are as spurious as the traditional square of opposition.

Hintikka and Quine once again supply solutions. Hintikka's modified rules by themselves constitute a solution. According to these (15) could not be derived as a conclusion unless '$(\exists x)(x = \text{Pegasus})$', i.e., 'Pegasus exists', were true. Similarly (17) would follow only if '$(\exists x)(x = \text{Pegasus})$', which is evidently false on the existential reading, were true.

In his *Mathematical Logic*, Quine's approach to introducing individual constants (even vacuous ones) is to treat all names as notational abbreviations for other expressions.[11] His more widely known use of

Russell's theory of descriptions to define away names, e.g., where 'Pegasus' is treated either as 'the one winged horse of Bellerophon' or as 'the one individual that pegasizes', is not found in *Mathematical Logic*. Here he uses descriptions to define away names, but the descriptions are then themselves analyzed in terms of class abstraction. The sentence 'Europe is a continent' becomes 'The one x which is Europe is a continent'. Using 'C' for 'is a continent', 'E' for 'is Europe', and the iota operator, this can be expressed as '$C \imath xEx$'. The description is in turn defined by means of class abstraction:

$$\ulcorner \imath xEx \urcorner \text{ for } \ulcorner \hat{z}(\exists y)(z \in y.(x)(x=y \cdot \equiv Ex)) \urcorner \text{ by D17}$$

All names are thus abstracts.

All the *names* we ever need, for pure logic and mathematics or for any other sort of discourse, are adequately provided by abstraction. For, suppose there is a name, say 'Europe', which is not defined as an abbreviation of an abstract but is instead simply a primitive term of geography. By a trivial revision of our geographical primitives and definitions we can reconstrue this name as an abstract, and in particular as a description (which is a special sort of abstract, by D17).[12]

Here individuals are identical with their unit classes. The problem of improper singular terms, i.e., those which are vacuous or not uniquely satisfied, is solved in a way similar to one suggested by Frege, which has aptly been called the "chosen object theory" because it has some object serve as the designatum of the improper description. For Quine, terms such as 'Pegasus' and 'the author of *Principia*', when defined as above, turn out to be abstracts which designate the null class.

With names available as disguised abstracts Quine offers metatheorems for their introduction.[13]

*231   If ϕ is like ψ except for containing free occurrences of η wherever ϕ contains free occurrences of a, then $\vdash$ $\ulcorner (a)$ $\phi \supset \psi \urcorner$.

*232   If ψ is like ϕ except for containing free occurrences of η wherever ϕ contains free occurrences of a, then $\vdash$ $\ulcorner \psi \supset (\exists a)\phi \urcorner$.

Quine does not apply either his definition of names or these principles to the troublesome inference 'Pegasus is a flying horse. therefore $(\exists x)(x$ is a flying horse)', but if we do so, we find that the premise would be false, because Pegasus turns out to be identical with the null class and as such is not a member of the set of flying horses. Universal instantiation is guaranteed to be truth preserving because the constant introduced as the instance will always have a designation.

It is instructive to digress here and compare Quine's approach with Hintikka's. For Hintikka, both senses of a free logic are achieved at once in his modified rules, which can make us lose sight of the independence of the two senses. In contrast, Quine modifies a single axiom to take care of the empty domain and then independently of this problem appeals to an analysis of names and descriptions as sanctioning the use of vacuous terms. In fact, his principles *103, *231, *134, and *232 are analogous to the traditional unmodified instantiation and generalization rules. This modification of a system to achieve a free logic for existence sentences while still leaving the question of empty terms unresolved is an important precedent for some of our later results concerning the non-existential readings of the quantifiers.

B.   Arguments Purporting to Establish
That We Don't Need a Free Logic

William Kneale and L. J. Cohen have questioned the need for a free logic in one of the senses of this term. Kneale argues that the requirement that the laws of logic hold in the empty domain is unnecessary.[14] He acknowledges that the unmodified rules of logic sanction the following reasoning:

$$\frac{\begin{array}{l}(x)Fx \supset Fy\\ Fy \supset (\exists x)Fx\end{array}}{(x)Fx \supset (\exists x)Fx}$$

Since he wishes to regard '$(x)Fx \supset (\exists x)Fx$' as a theorem even though it is false for the empty domain, Kneale concludes that we should confine ourselves to non-empty domains and dismiss the empty one from consideration. He asserts that there is no loss in thus confining ourselves, and to bolster this view offers an argument purporting to show that there is no empty domain of individuals. His reasoning is that

(23) Individuals are *arguments* for lowest level propositional functions.

(24) Propositional functions specify your domain.

(25) There can be no propositional function without an *argument*.

Therefore

(26) There can be no domains without individuals.

I have emphasized the word 'argument(s)' because I believe there is an equivocation in this demonstration. In (23), 'argument' is used to refer to something non-linguistic, viz., an individual; (25) is plausible if 'argument' is taken as a linguistic entity, viz., the variables in propositional functions or one of their substituends. But if we construe (23) and (25) in this way, the conclusion does not follow. On the other hand if 'argument' in (25) is taken as an individual or non-linguistic object, i.e., argument as that which satisfies a propositional function, then it is false. Many propositional functions, e.g., '*x* is a unicorn' are not satisfied by an object.

Commenting on what he takes to be the motivation for having a free logic, Kneale says it stems from a dogma of empiricism that no existential propositions can be established *a priori*. He is right in thinking this is not a good motive, but wrong in think that it is *the* motive. The requirement that logic be free of special assumptions ought not to be rooted in such questions of epistemology as how we acquire knowledge in logic or mathematics, but in the generality of the laws of logic. These laws can be described as containing only variables in addition to the logical constants. The motivation for having a free logic is that any meaningful substitution for the variables should yield a truth. Analogously, the rules of inference should be truth-preserving. Similar reasoning applies to arguments that a free logic is motivated by the belief that no existential sentences are analytic. The need for a free logic is independent of the issues of whether logic is established *a priori* or *a posteriori* or whether the truths of logic are analytic or synthetic.

L. J. Cohen claims that "the notion of an empty domain of discourse is self-contradictory."[15]   He says that existence means class intersection or class membership. For example

'Men exist'   means   The class of men intersects with
                                        some class such as that of mammals.

'Socrates exists'   means   Socrates is a member of the class of men.

Non-existence is the absence of class intersection or class membership.

'Flying horses don't exist'   means   The class of flying horses and the class of physical objects (for example) do not intersect.

'Pegasus doesn't exist'   means   Pegasus is not a member of a class.

Concentrating on the fact that at different times and in different contexts a term's extension might be said to change, Cohen claims that non-existence is always relative to some domain (class) whose existence in that context is unquestioned. According to this account, one does not raise questions of the emptiness or non-emptiness of domains, for to do so is no longer to regard them as your domains.[16] All of these remarks, however, simply obscure the fact that a domain is itself a class and the conjecture that there is an empty domain is simply that a class might have no members. It seems both arbitrary and false to say that one cannot inquire whether a domain does or does not have members.

C.   Semantical Frameworks for the Existential Reading

A first order language can be thought of as an ordered quadruple $\langle P, V, L, S \rangle$ where P is the class of predicates, V is the class of individual expressions, i.e., individual constants and individual variables, L is the class of logical constants and parentheses, and S the class of WFFs generated by the usual definitions. Valuations (val) are functions assigning T or F (truth or falsity) to the WFFs 'A', 'B', etc. In this framework the variables are assigned a domain D of objects. Predicates are assigned sets of objects from the domain in such a way that predicates of degree n are assigned n-tuples where $n \geqslant 0$. Our first order language will contain individual constants. To the individual constants we assign the individuals of the domain. We adopt here 't_1'...'t_n' as meta-linguistic variables having individual constants as their values and 'P' as a meta-linguistic variable for predicates. The following are truth conditions for the existential reading which are suitable to the unmodified rules.

val $(Pt_1,...,t_n) = $ T iff $\langle$val $(t_1),..., $ val $(t_n)\rangle \in$ val (P)

val $(\sim A) = $ T iff val $(A) = F$

val $(A \& B) = $ T iff val $(A) = $ val $(B) = $ T

val $((x)A) = $ T iff $(d)(d \in D \supset $ val $(d/xA) = $ T)
       wherever x is free in A

val $((\exists x)A) = $ T iff $(\exists d)(d \in D \& $ val $(d/xA) = $ T)
       wherever x is free in A[17]

Let us take a quick survey of some other ways, e.g. Tarski's, Mates', and a substitutional account, of providing a semantics for the existential reading. In all of these the crucial question centers on the condition given for quantified sentences. Tarski speaks of open sentences being satisfied by sequences of objects. A sequence satisfies an open sentence of the form of a universal quantification if and only if, no matter how we vary the sequence with respect to the variable bound by the quantifier, the open sentence is still satisfied. The reader should note that Tarski's account of quantification is essentially non-substitutional, since there is no mention whatever of instances but only of sequences of objects.

Another account, and one which appeals to instances in the truth conditions for the quantifiers, is found in Benson Mates' *Elementary Logic*. A universal generalization $(x)(A)$ is true if and only if every β-variant of A is true. A β-variant is a substituend which can be assigned different objects. Thus one β-variant of 'Socrates' would be the assignment Socrates, another β-variant would assign Plato to 'Socrates', and so on. The basic idea is to reinterpret one constant so that different objects from the domain are assigned to that constant. For the universal generalization every object will be assigned to the constant in question. Roughly speaking, $(x)(A)$ is true if and only if an instance of it is true and remains true no matter what objects are assigned to the substituend of that instance. This is somewhat analogous to Tarski's method. Tarski varies objects in sequences without using constants, while Mates varies assignments of objects to an instance. If, like the present author, one is inclined to regard an account of the quantifiers as substitutional when instances are appealed to, then Mates' method is in this crucial respect "substitutional". The method of β-variants differs, though, from a standard substitutional account with referential force. For one thing Mates' method doesn't require that every object be named, but only that there be at least one constant that can be continually reinterpreted to refer to whatever objects there are.

Yet another possibility is to design a standard substitutional account so that it can serve the existential reading. This is easily enough accomplished by using the conditions given above for atomic sentences, negation, and conjunction, and then supply as our generalization clause the following:

val $((x)A) = T$ iff for all individual constants t of the language V, val $(t/xA) = T$.

This account is both substitutional and referential. It is substitutional because of the appeal to all (or some) of the instances. It is referential because the instances will have the referential force given to the atomic sentences.

None of the frameworks mentioned so far in this section is suitable to some adaptations of the Hintikka-type modified rules. Part of the utility of these modified rules is that they allow one to consider 'Pegasus is a flying horse' as true while treating '$(\exists x)(x$ is a flying horse)' as false. According to the above truth conditions, either 'Pegasus' would not be assigned an individual, so that 'Pegasus is a flying horse' would be false, or, if it were assigned an individual, then '$(\exists x)(x$ is a flying horse)' would be true and so not conform to its intended existential reading.

The problem is thus one of adopting a semantics in which variables have values in a given domain when it is precisely this membership in a domain which provides the rationale for reading '$\exists x$' as 'there exists'. In the interests of some varieties of free logic, we want to allow atomic sentences with vacuous individual constants to be true in some cases and not in others. For instance, the above Pegasus sentence is true but 'Pegasus is a flying fish' is not.

Hughes Leblanc has shown how we might construct a semantical system which allows us to sketch a solution.[18] He has adopted, in addition to an ordinary domain which contains the values of the variables, another domain that contains objects assigned to constants, constants which are vacuous with respect to the first domain. These are called inner and outer domains: D_i and D_o. 'Pegasus is a flying horse' would be true because 'Pegasus' is assigned an object from the outer domain and this object is found among a set of horses which itself is a subset of the union of the inner and outer domains. Yet '$(\exists x)Fx$', i.e., 'There exists a flying horse', does not follow because '$(\exists x)(x = Pegasus)$' is not true with respect to the inner domain. Leblanc's outer domain solution is reminiscent of Frege's chosen-object theory. For Frege a vacuous term had to be assigned a denotation, i.e., an object, in order for the sentence in which it occurs to have a truth value. Leblanc has improved upon this by providing a domain in which not only do we have the chosen object, but it is chosen so that we get just the true and false vacuous atomic sentences we want.

To accomplish this we can adopt the following truth conditions.

val $(Pt_1,...,t_n)$ = T iff ⟨val (t_1),..., val (t_n)⟩ ε val (P)
                       where P and $t_1,..., t_n$ have their valuation with respect to both the inner and the outer domain.

val $(\sim A)$ = T iff val (A) = F

val $(A \ \& \ B)$ = T iff val (A) = val (B) = T

val $(\ (x)A)$ = T iff $(d)(d \epsilon D_i \supset$ val (d/xA) = T)
                 wherever x is free in A.

val $(\ (\exists x)A)$ = T iff $(\exists d)(d \epsilon D_i \ \&$ val (d/xA) = T)
                 wherever x is free in A

Another solution is provided by Hintikka himself.[19]   Hintikka's semantics of model sets leaves aside the question of assigning objects to the individual constants and predicates of the atomic sentences. The only condition for an atomic sentence is that it be a member of a model set, and this is accomplished if not both the sentence and its negation are members. Thus 'Pegasus is a flying horse' may be a member if its negation is not. Presumably 'Pegasus is a flying fish' is not a member of a given model set, because its negation is a member. The condition for an existential generalization's being a member of a model set is that one instance as well as a relevant existence sentence be a member of the set. Thus

If $(\exists x)Fx$ is a member of the model set, then Ft is a member too and so is $(\exists x)(x = t)$

Hintikka's idea is to allow one to generalize existentially only on those formulas that have values for their variables. It is as though the atomic sentences are divided into those that are vacuous and those that are not. Sentences of the form $(\exists x)(x = t)$' accompany the non-vacuous atomic sentences and permit the existential generalization.

There has been some confusion as to whether Hintikka's quantifiers are substitutional or referential.[20] If by "substitutional" we mean that the quantifiers are explained in terms of instances, then to this extent his quantifiers are substitutional. However, Hintikka's quantifiers apply only when the substituends refer. So his quantifiers also qualify as being referential. Since Hintikka's use of $(\exists x)$' is existential, the values of his variables are restricted to existents.

Hintikka has, more than most authors of similar persuasion, put his modified rules and semantics to interesting philosophic uses. He has questioned important existential conclusions that Moore and Descartes drew. Moore argued that hands exist ($(\exists x)(Hx)$) because this is a hand (Ht). Descartes argued that I am ($(\exists x)(x = i)$ because I think (Ti). Hintikka remarks that these inferences are not correct unless one has additional existential premises.

What is common to all the frameworks for the existential reading is that the semantics are referential, either as in Tarski's case via the satisfaction relation which is part of the conditions for the quantifiers, or as in the substitutional semantics via the referential force of the instances. Put somewhat differently, the existential reading requires the concept of a domain in its semantics. Furthermore these domains used to explicate the "existential quantifier" are populated only by existents.

Part II The Non-Existential Readings

D.   Do We Need a Free Logic for the Non-Existential Readings of the Quantifier?

I have so far discussed the need for a logic free of existence assumptions, and have taken for granted that existence is expressed by means of the particular quantifier. Quite naturally, the question arises: if we stop reading the quantifier as having existential import, does the need for a free logic disappear? This is actually two questions, because there are two senses in which a logic is free. I will first explain why it is required that no theorems of logic should be particular generalizations. The explanation given here is informal and will be supplemented by some further discussion after we consider the semantics for the non-existential readings. With regard to the second sense in which logic should be free—that it should apply to any descriptive constants—we will find that the unmodified rules suffice as they apply to descriptive constants. We could adopt Quine's system from *Mathematical Logic* to the non-existential readings. We would borrow that part of Quine's system which blocks proving particular generalizations as theorems, but dispense with his treatment of names.

Leibniz's dictum that the theses of logic be true in all possible worlds is usually applied to possible domains of objects. Unless understood differently, this idea is inappropriate to a substitutional reading of the quantifiers. In substitutional quantification we deal with classes of substituends and not necessarily with domains of objects. Later on I will consider cases where substitutional quantification is not referential and where the truth conditions do not mention domains.

Leibniz's point about logical truths, however, can be stated without reference to domains. The theses of logic are sentences which contain logical constants: negation, conjunction, one or both of the quantifiers, notational devices like parentheses, and then only variables. As Russell put it, they are "propositions which can be expressed in the language of pure variables."[21] The significance of describing the form of a logical truth is that we can arrive at a definition of such truths which makes mention not of domains but only of constants put in place of the variables. A logical truth is a sentence containing only logical constants, notational devices, and variables, and which has the property of being true no matter what constants (so long as they are grammatically appropriate) are substituted for the variables. This account allows us to indicate to some extent why, even on the substitutional view, particular generalizations should not be considered theses. The truths of logic are absolutely general and should hold no matter whether any or even all less general sentences are true or false. We might arrange sentences according to a series from the general to the more specific. This could be done by logical form, e.g., universal sentences before particular sentences and particular sentences before singular ones. Logic should not involve assumptions about any less general truths; in particular, it should not commit us to the truth of even one singular sentence. Logic ought to leave open the possibility that no singular sentence is true. Consider '$(\exists x)(x = x)$' read as 'It is sometimes true that things are self identical.' The reason why we should reject this as a theorem even on the substitutional approach is that it involves an assumption. The assumption is not one of existence but of the truth of at least one sentence. It precludes the possibility that every singular sentence of the form $a = a$ is not true. It is because we want logic to be completely neutral as to special truths and falsehoods that we take this possibility seriously. Lambert speaks of a "free logic" as a logic free of existence assumptions. For non-existential readings of the quantifier, and especially for non-referential substitutional accounts, the term "free logic" must be given an extended meaning.

In the second sense in which a logic should be free—that the rules of inference apply to every putatively meaningful constant—the substitutional reading is already a free logic and there is no need to revise the ordinary rules of quantification. Recall Marcus's treatment of the apparent counter example to generalization.

Pegasus is a flying horse.
It is sometimes true that horses fly.

Here the conclusion is plausibly true if the premise is. Thus there is no motivation to adopt Hintikka-type modified rules, Quine's variant of the chosen object theory, or even Russell's theory of definite descriptions. (One may indeed speculate as to the uses of the theory of definite description when the quantifiers are not read existentially.)

In Chapter Two additional claims were made for the divergent substitutional reading. The non-existential reading was said to avoid the difficulties associated with the existential construals of

$$\Box (\exists y)(x = y)$$
$$(x)\,\Box (\exists y)(x = y)$$
$$\Box\,(x)(\exists y)(x = y)$$

However, none of these formulas are provable as theorems in a free logic, on either the existential or the non-existential construal. What of the Barcan formula,

$$\Diamond (\exists x)Fx \supset (\exists x)\Diamond Fx \qquad ?$$

Does a non-existential reading save it from criticism? The answer is that even with an non-referential substitutional semantics this formula remains controversial.   In his paper "Semantical Considerations of Modal Logic," Kripke, using a referential semantics suitable to the existential reading, constructed counterexamples to the formula.   Nonreferential analogues for these Kripkean domain-oriented semantics have been constructed by Dunn and Leblanc.[22]   Thus there are reasons to believe that there are counterexamples to the Barcan formula even if we dispense with referential semantics.   Prof. Marcus continues to defend this principle but along somewhat different lines.[23]

What of substitutional quantification and intensional contexts generally?   Is the substitutional generalization from

$\Box(9>7)$

to

$(\exists x)\Box(x>7)$

permissible or are such cases of quantifying unintelligible as Quine originally maintained?   Kripke in his paper "Is There a Problem about Substitutional Quantification?" has substantiated the dramatically different role of substitutional quantification in opaque contexts.

the intelligibility of substitutional quantification into a belief or modal context is guaranteed provided the belief or modality is intelligible when applied to a closed sentence.   The reason is that, in the theory of substitutional quantification . . . , the truth conditions of closed sentences always reduce to conditions on other closed sentences.[24]

Perhaps a moral to be drawn from this is that quantification *per se* is not essential to the problems of modalities and belief contexts.

Semantics for Substitutional Readings

I take an approach here that relies on Carnap's method of state descriptions.   Recall that our first order language is the ordered quadruple $\langle P, V, L, S \rangle$.   A state description is the class of atomic WFFs belonging to S (S is the set of WFFs of the language) or their negations, but not both.   There are many state descriptions and Carnap thought of each of them as describing a different Leibnizian possible world; one among these describes the actual world.   Carnap supplemented this account of state descriptions with semantical rules of designation which assigned objects to the predicates and individual constants of the atomic WFFs, but since I want to develop a non-referential semantics for the substitutional reading I shall forego such talk of worlds and the assignment of objects. In referential semantics, relations such as designation, denotation, or satisfaction are semantical primitives. We adopt a semantical notion which will not involve objects. An obvious choice is to take the predicate 'is true' as the touchstone for our semantics. We speak of either individual atomic WFFs or of classes of them such as state descriptions as true. I will use 'S_t' as a name for the true state description in the following conditions.

val $(A) = \text{T}$ iff $A \in S_t$;

val $(\sim A) = \text{T}$ iff $A \notin S_t$; or val $(A) = \text{F}$

val $(A \, \& \, B) = \text{T}$ iff $A \in S_t$ and $B \in S_t$; val $(A) = $ val $(B) = \text{T}$

val $((x)A) = \text{T}$ iff for individual constants of the language s/xA
$\quad \in S_t$; or val $(s/xA) = \text{T}$ wherever x is free in A.

val $((\exists x)A) = \text{T}$ iff $(\exists s)(s \in \text{V} \, \& \, s/xA \in S_t)$ or $(\exists s)(s \in \text{V} \, \& \,$ val
$\quad (s/xA) = T)$ wherever x is free in A and V is the
$\quad$ class of individual expressions of the language.

Kripke's recent paper on substitutional quantification puts the semantics of this subject on the firmest of grounds.

> There never was any problem about substitutional quantification. For any class of expressions C of a language L_0, we can extend the language by adding substitutional quantification with this class of expressions as the substitution class, whether or not some, all, or none of the expressions in the class denote .... The issue of whether truth conditions have been given for substitutional languages is one of mathematical fact, not philosophical opinion.... Any argument to the contrary, therefore, must be fallacious: it puts its proponents in the camp of the circle squarers and the angle trisectors.[25]

I shall try to sketch Kripke's approach and then will compare it with the one taken above.

Kripke provides a restricted language L_0 and at first assumes that the truth conditions for its sentences have been given (later in the essay he provides these conditions using a new and rather unique semantical primitive). Probably the most familiar sentences that could be included in L_0 are the atomic sentences which occur in the first clause of our other constructions as well.) L_0 is extented to the language L by syntactical rules which introduce negations, conjunctions, and quantifications on the sentences of L_0. The truth conditions for negations and conjunctions are the usual ones and the conditions for quantification are substitutional.

While L_0 provides the substituends for the substitutional quantifiers of L the question of whether these substituends designate/refer is left open. If the sentences of L_0 are atomic sentences the individual constants need not be assigned objects. Moreover the sentences of L_0 need not be our atomic sentences and the substituends from L_0 may be predicate constants or sentences, and these presumably need not be assigned objects. L_0 may even contain opaque constructions. It may also contain negations, conjunctions and quantifications so long as we distinguish these from the negations, conjunctions and quantifications of L. By setting up and separating L_0 and L in this way, Kripke is able to establish results for substitutional quantification in all its varied uses, referentially and non-referentially, in opaque and in transparent constructions, and quantifying over non-nominal positions.

In this monograph I have maintained that substitutional quantification pertains to how generalizations are true (truth in all or in some instances) and that the way in which instances acquire their truth value is left open. Kripke has much more adequately made the same point about the independence of the truth conditions for quantifiers and sentences of the language L, and truth conditions for the sentences of the language L_0. This relation of Kripke's apporach to the one I have adopted is as follows:

Truth Conditions for

$$
L \begin{cases} L_0 \\ \text{Negations} \\ \text{Conjunctions} \\ \text{Quantifications} \end{cases} \quad \text{Instances} \quad \begin{cases} \text{Atomic sentences} \\ \text{Negations} \\ \text{Conjunctions} \\ \text{Quantifications} \end{cases}
$$

Should my talk of instances prove unrewarding, the indulgent reader is requested to translate my remarks into Kripke's terminology.

A Semantics for the Neutral Reading

The following are semantical frameworks for the neutral reading. A guiding consideration in formulating these is that the particular generalization '$(\exists x)(x$ is a flying horse)' could be true and follow from 'Pegasus is a flying horse' which is also true. A first suggestion is simply to adopt the semantical frameworks employed for the existential reading but with one difference. Let our domains be populated with beings as well as existents, and let the quantifiers bind variables that have beings and not merely existents as their values.

A second semantical framework utilizes the device of inner and outer domains (i.e., D_i, D_o).

val $(Pt_1,...,t_n) = T$ iff $\langle$val $(t_1),...,$ val $(t_n)\rangle \epsilon$ val (P) where P and $t_1,...,t_n$ have their valuation with respect to both the inner and the outer domain.

val $(\sim A) = T$ iff val $(A) = F$

val $(A \ \& \ B) = T$ iff val $(A) =$ val $(B) = T$

val $((x)A) = T$ iff $(d)(d \epsilon D_i \cup D_o \supset$ val$(d/xA) = T)$ wherever x is free in A

val $((\exists x)A) = T$ iff $(\exists d)(d \epsilon D_i \cup D_o \ \& $ val $(d/x) = T)$

A third framework for the "there is" reading is a hybrid. It combines domains and the device of state description which is employed for non-referential substitutional quantification.

val $(Pt_1,...,t_n) = T$ iff $\langle$val $(t_1),...,$ val $(t_n)\rangle \epsilon$ val (P) or $A = Pt_1,...,t_n$ $\& A \epsilon S_t$

val $(\sim A) = T$ iff val $(A) = F$

val $(A \ \& \ B) = T$ iff val $(A) =$ val $(B) = T$

val $((x)A) = T$ iff $(d)(d \epsilon D \supset$ val $(d/xA) = T)$ or $(s)(s \epsilon V \supset$ val$(s/xA) = T)$ wherever x is free in A

val $((\exists x)a) = T$ iff $(\exists d)(d \epsilon D \ \& $ val $(d/xA) = T)$ or $(\exists s)(s \epsilon V \ \& $ val $(s/xA) = T)$ wherever x is free in A

In Chapter Two the question was raised whether one could have a non-Meinongian account of the 'there is' locution. If we restrict the word "Meinongian" to views which posit non-existent beings, then there is a sense in which the first two semantical frameworks for the neutral reading are "Meinongian" and the last is not. Having domains which include non-existent beings and quantifying over them appears to represent a more extreme view than merely accepting as true certain vacuous sentences (such as the Pegasus sentence above) and then substitutionally quantifying over them. The less extreme view though, does make the neutral reading depend on a substitutional semantics of the non-referential sort.

I should like to try to formulate an argument against adopting a Meinongian semantics, and for the substitutionally oriented account of the neutral reading. The parties in this dispute all treat some sentences about non-existent objects as true and allow generalization with regard to them. There appears to something *ad hoc* about positing non-existent beings. We posit them simply because we wish to consider certain sentences true. Meinongian entities are in some respects like shadows. A person's body accounts for his shadow and not vice-versa. Analogously the truth of some vacuous sentences motivates introducing non-existent beings, but these beings don't appear to provide independent grounds on their own for regarding sentences as true. Candor and simplicity suggest that if we want to treat the sentences in question as true then we do just that and nothing more.

E. Informal Comments on the Above Frameworks

In referential semantics like those provided above for the existential and the neutral readings, assignments are made with respect to domains of objects. There is a variety of methods for interpreting or making assignments to predicates and individual constants; Tarski spoke of satisfaction, Carnap made use of designation, and Martin described multiple denotation. Satisfaction is a many-one relation between sequences of objects in the domain and sentential functions. Carnap's designation is a one-one relation among predicates, individual constants and sets and individuals of the domain. Multiple denotation is a one-many relation between predicates and individuals in the domain. The following rule of denotation will give us a better idea of what these relations are like.

$(x)(a$ Den x iff $--- x ---)$

Here 'a' can be replaced by the name of a predicate such as 'p' while '――― x ―――' becomes a sentential function like 'Px'. Thus for example, if the sentence " 'P' Den x" were assumed true, we could derive the sentential function 'Px'. Of some interest in the fact that, given the unmodified rule of generalization, " 'P' denotes x" would entail '$(\exists x)$ Px'. If construed existentially, this would mean that denoting entails existence. If this consequence is undesirable, one might adopt a non-existential reading.

Perhaps I should comment on the differences between the semantics for the existential reading with regard to the Hintikka-type modified rules and the semantics for the neutral reading. In the former we want to consider the Pegasus sentence true but the generalization false and in the latter we want both sentences to be true. Leblanc, as we saw, provided a semantics for this existential reading by adopting

val $((\exists x)A)$ = T iff $(\exists d)(d \in D_i$ & val (d/xA) = T)

as his truth condition for an existential generalization. He apparently wanted to exclude

val $((\exists x)A)$ = T iff $(\exists d)(d \in D_i \cup D_o$ & val (d/xA) = T).

However, I adopted this very condition to render a neutral reading for '$(\exists x)$' i.e., 'there is'. Leblanc presents an interesting, if not fully convincing, argument as to why we should not allow such generalizations. He considers the case in which 'a' is a vacuous singular term having no value and 'F' stands for 'does not exist'. 'Fa' would be read as 'a does not exist'.

> Under this interpretation of 'a' and 'F'; the existential quantification '$(\exists x)Fx$ is clearly false even though the replacement of 'x' everywhere in 'Fx' by the parameter 'a' is true.[26]

But '$(\exists x)Fx$' is false only if we persist in reading '$\exists x$' as 'there exists'. Is there any purely logical reason why we must do so? Furthermore are there any philosophical or logical reasons to modify our rules of inference or semantical conditions, other than that of wishing to preserve a connection between '$\exists x$' and existence? In the absence of the above we can equally well claim that '$\exists x$' be read as 'there is'.

The semantics I provided for the non-existential substitutional reading relied on Carnap's state descriptions. Other authors have given accounts which might have served equally well. Jaakko Hintikka revised the notion of a state description into his own model sets, but he does not use model sets to justify a non-existential substitutional reading.[27] Henry Hiz has presented a semantical theory which could also be adopted as a framework for this reading.[28] It goes much further than the above account in trying to sketch a semantics for natural languages as well. It is called 'aletheic semantic theory' because truth is its primitive notion.

Leblanc, too, has done a great deal of work to show that the substitutional view can be justified.[29] He has considered the possibilities of a truth-value semantics in which we compute the truth value of quantified formulas in terms of subformulas. In some of his accounts, such as those mentioned earlier, Leblanc has shown an interest in retaining the concept of existence, and in so doing assumes some referential, i.e., domain-oriented, account of the truth of atomic sentences. In other works, which are perhaps more representative, Leblanc dispenses with domains.

Richard Martin has also provided a number of semantical frameworks which could be of use here.[30] Of most interest to us is a primitive which he calls comprehension. Roughly speaking, the idea is that one expression comprehends another. Martin begins with comprehension for predicate constants and then introduces it for individual constants. In a statement of the form 'a comprehends b', 'a' and 'b' are names of constants which could occur in some object language. The semantical language used is said to be 'non-translational' because, although it contains names of expressions of the object language, it does not contain the object language expressions themselves. This allows us to refrain from assigning objects to the terms, if we so choose. Although Martin does not (and would not) do so, one could adopt the notion of comprehension to assign a value to atomic sentences:

$$\text{val } (Ps_1 \dots s_n) = T \text{ iff P comprehends } \langle s_1 \dots s_n \rangle$$

We could then lay down the remaining truth conditions so that they suited the substitutional reading.

In his study of substitutional quantification Kripke proposes a semantic primitive that is designed to accommodate what I have called non-referential substitutional quantification. At issue are the truth conditions for the sentences of his restricted language L_0. For our present purposes we can think of L_0 as made up of atomic sentences. Kripke's semantic primitive is the two placed predicate Q. The Q predicate is modeled on the concept of quotation. An example of a Q sentence is

Q (Cicero, $\overline{\text{Cicero}}$)

(where $\underline{\quad}$ forms a structural descriptive i.e., quotationlike, name of $\underline{\quad}$. The Q relation is not to be confused with the denotation relation because Kripke specifically wishes to leave undecided the question of whether the terms of the language L_0 (the individual constants in our atomic sentences) denote.

It remains for us to constrast the semantics for a non-existential substitutional reading and that for the existential and Meinongian readings. The semantics for the latter readings makes use of the notion of a domain. A model for these quantificational systems is an ordered pair $\langle f, D \rangle$ where D constitutes the domain of the model and f the interpretation function, assigning subsets of D to each of the predicates and individuals from D to each of the individual constants. The truth of atomic sentences is relative to the assignments of individuals and sets to the individual and predicate constants in that sentence. The truth of a quantified sentence is also dependent (directly or indirectly) on that domain. For the non-existential substitutional approach, our semantics makes no mention of a domain. If we tried to construct a parallel for the above notion of a model we would get an ordered pair $\langle f, S \rangle$ where S is the set of sentences of the language and f an interpretation function assigning some value to the sentences of S. In our use of the method of state descriptions, f assigns truth to a state description constituting a subset of S; our atomic sentences are then true relative to that assignment. The quantified sentences are in turn true relative to the assignments made to the atomic sentences. Some might regard it as misleading to call $\langle f, S \rangle$ a model at all. The terms 'model', 'model theory', 'model theoretic', etc. are often thought of solely in connection with domain-oriented semantics. Indeed, the above distinction between referential and non-referential semantics is often conveyed simply by speaking of the former as model theoretic.

F. A Second Look at Quantification and the Empty Domain

The most precise argument for considering universal sentences true in the empty domain and particular ones false is found in a paper by Mostowski.[31] He relies on a variant of the following truth condition.

$$\text{val } ((x)A) = \text{T iff } (d)(d \in D \supset \text{val } (d/xA) = \text{T})$$

For the case of the empty domain, the antecedent '$d \in D$' would be false and so the second component of the biconditional would be vacuously true, which in turn would make the right component—the universal sentence—true. By similar reasoning, particular quantifications are false in the case of the empty domain.

$$\text{val } ((\exists x)A) = \text{T iff } (\exists d)(d \in D \text{ \& val } (d/xA) = \text{T})$$

If '$d \in D$' is false, the second component of the biconditional would be false, and hence the first component would be false too. In so far as the neutral or substitutional readings rely on a domain, the same argument would apply. Is it possible, though, to reconstruct such an argument for a non-referential view? Consider our formulation of the truth condition for such a universal quantification.

$$\text{val } ((x)A) = \text{T iff } (s)(s \in V \supset \text{val } (s/xA) = \text{T})$$

If $s \in V$ were false, i.e. there were no substituends, then by reasoning similar to the above the universal conditional would be true and a particular generalization false.

By providing the appropriate systems and semantics, we have justified three readings of the particular quantifier. Of especial importance for the purpose of this monograph are the foundations provided for the non-existential readings. We can now confidently employ the neutral and substitutional readings in the remaining chapters to evaluate the various claims made about existence and quantification.

Notes

1. R. M. Martin, "Existential Quantification and the Regimentation of Ordinary Language," *Mind*, 71 (October, 1962), p. 258.
2. W. V. Quine, *Mathematical Logic* (New York: Harper Torchbooks, 1951), p. 88.

3. W. V. Quine, *Selected Logic Papers* (New York: Random House, 1966), pp. 221-222.

4. See the following: C. Lejewski, "Logic and Existence," cited in *Logic and Philosophy*, ed. G. Iseminger (New York: Appleton-Century-Crofts, 1968), pp. 167-181 (hereinafter cited as Lejewski, "Logic and Existence"); K. Lambert, "Free Logic and the Concept of Existence," *Notre Dame Journal of Formal Logic*, 8 (April, 1967), pp. 133-144; J. Hintikka, "Existential Presuppositions and Existential Commitments," *Journal of Philosophy* (1959), p. 133.

5. In this note we will examine some formulations of quantification theory, by commenting on the role of constants, different kinds of variables, unspecified individual constants, arbitrary individuals, schematic letters, and free and bound variables.

To begin with consider the following attempt to state the rule of universal instantiation.

for variables
$$\frac{(x)Fx}{Fy}$$

for terms
$$\frac{(x)Fx}{Fa}$$

It is not at all clear what this asserts. One way of opening this Pandora's box is to note that 'x' and 'y' are usually used as object language variables while 'F' and 'a' are frequently construed as abbreviations for specific predicate and individual constants, e.g., 'F' for 'is fat', 'a' for 'Albert'. But then $\frac{(x)Fx}{Fa}$ is not general enough to do the job we intended, since it is merely one particular argument. If we use Quine's meta-linguistic variables, our statement of the rules is not open to this objection.

Let us make some further distinctions beginning with that between a variable and a constant. A variable either ranges over a domain of objects which serve as the values of the variable or has a range of expressions which serve as the substituends of the variable. A constant, or more specifically an individual constant, is paradigmatically a name standing in a one-one relation to its nominatum. A predicate constant can be construed as being in a one-one relation to some set. In Quine's case, meta-linguistic variables have object-language expressions as their values. Although we have a clear idea of variables, constants and meta-linguistic variables, we are at a loss to explain precisely what an unspecified individual constant is; 'a' is supposed to play this sort of role in some formulations of quantification theory. However, the onus remains on those who use 'a' in this way to provide some sort of satisfactory analysis. Other authors have made reference to arbitrary individuals which are distinct from unspecified individual constants. These, too are in need of further analysis. Quine speaks of schematic letters (placeholders, dummy letters, slots) which are neither variables of any sort or level, not constants. It is not clear what these schematic letters are; I shall have more to say on this subject in the last chapter.

Yet another distinction is that between free and bound variables. An *occurrence of a variable* is bound if it occurs in the scope of a quantifier, otherwise it is free. A *variable* is bound if one of its occurrences is bound, and free if one of its occurrences is free. Greater care can be shown in the construction of a system of logic if we prohibit formulas with free variables (propositional functions) from being lines in a deduction and thus also from being theorems. This has more often been recognized as an ideal than carried out in practice. Free variables frequently play two different roles,—as surrogates for universally quantified sentences and as individual constants.

On the one hand Church (*Introduction to Mathematical Logic*, Princeton: Princeton University Press, 1956, Vol. 1, p. 46) repeats a point made by Russell in the second introduction to *Principia*, that formulas with free variables can be dispensed with in favor of universally quantified formulas. For example, consider '$(x)Fx \supset Fy$'. One could as well have proved '$(y)((x)Fx \supset Fy)$'. Just so, free variables might just as well be bound by initial universal quantifiers.

On the other hand free variables sometimes play the role of constants. One case of this is found in texts which let the following lines (3) and (4) serve as the contradiction necessary for an indirect proof.

(1) $(x)Fx$	Given	
(2) $(\exists x)\sim Fx$	Given	
(3) $\sim Fy$	2	Existential Instantiation
(4) Fy	1	Universal Instantiation

Lines (3) and (4) are contradictory precisely on the condition that the free variable is taken as a constant.

Quine's system in *Mathematical Logic* is one in which propositional functions do not occur either as lines in deductions or as theorems. Both the traditional rules as well as Hintikka's are defective in this respect since 'Fx' could be a line in a proof and '$Fx \supset ((\exists y) (y = x) \supset Fy)$' is a theorem.

6. K. Lambert, "Free Logic and the Concept of Existence," *Notre Dame Journal of Formal Logic*, 8 (April, 1967), pp. 133-144.

7. Russell, *Introduction to Mathematical Philosophy*, p. 204.

8. C. Lejewski, "Logic and Existence," *Logic and Philosophy*, ed.

G. Iseminger (New York: Appleton-Century-Crofts, 1968), pp. 167-181.

9. For example see K. Lambert, "Free Logic and the Concept of Existence," *Notre Dame Journal of Formal Logic*, 8 (April, 1967), pp. 133-144.

10. J. Hintikka, "Existential Presuppositions and Existential Commitments," *Journal of Philosophy* (January, 1959), p. 133.

11. Quine, *Mathematical Logic*, pp. 146-152.

12. Ibid., p. 149.

13. Ibid., pp. 171-172.

14. William and Martha Kneale, *The Development of Logic* (Oxford: Clarendon Press, 1962), pp. 706-707.

15. L. Jonathan Cohen, *The Diversity of Meaning* (New York: Herder and Herder, 1963), p. 261 (hereinafter referred to as *The Diversity of Meaning*).

16. Ibid., pp. 263-264.

17. Actually though these truth conditions for quantified sentences are perspicuous, they are somewhat incorrect. The problem is that 'd' in these formulas is systematically ambiguous, i.e., in '$d \in D$', 'd' ranges over objects from the domain; while in 'val $(d/xA) = T$', 'd' ranges over expressions whose values are the objects in the domain. This inexact formulation does not affect any points made here or later about referential semantics. More exact formulations of this sort can be found in Robinson and in Kripke. In A. Robinson's *On the Mathematics of Algebra* (Amsterdam: North Holland Publishing Co., 1951), p. 19, we find the following:

Given any predicate $X(v)$ of $K(C)$,

2.7.6. $[(v)X(v)]$ holds in M if and only if $X(a)$ holds in M for all the object symbols a in L which correspond to objects in M.

2.7.7. $[(\exists v)X(v)]$ holds in M if and only if $X(a)$ holds in M for at least one object symbol a in L which corresponds to an object in M.

In S. Kripke's "Semantical Considerations on Modal Logic," in L. Linsky, ed., *Reference and Modality* (London: Oxford University Press, 1971), p. 67, these conditions are expressed this way.

Assume we have a formula $A(x, y_1, ..., y_n)$, where x and y_1 are the only free variables present, and that a truth-value $\phi(A(x, y_1, ..., y_n), H)$ has been defined for each assignment to the free variables of $A(x, y_1, ..., y_n)$. Then we define $\phi((x)A(x, y_1, ..., y_n), H) = T$ relative to an assignment of $b_1, ..., b_n$ to $y_1, ..., y_n$ (where the b_1 are elements of U), if $\phi((A(x, y_1, ..., y_n), H) = T$ for every assignment of

$a, b_1, ..., b_n$ to $x, y_1, ..., y_n$, respectively, where $a \in \psi(H)$; otherwise, $((x)A(x, y_1, ..., y_n), H) = F$ relative to the given assignment. Notice that the restriction $a \in \psi(H)$ means that in H, we quantify only over the objects actually existing in H.

Here H is a domain (for Kripke a possible world) and ϕ and ψ are valuations for that domain.

18. H. Leblanc and R. H. Thomason, "Completness Theorems for Some Presupposition-free Logics," *Fundamenta Mathematicae*, 62 (1968), pp. 125-164.

19. J. Hintikka, *Models for Modalities* (New York: Humanities Press, 1969), pp. 24-43.

20. D. Davidson and G. Harman, eds., *Semantics of Natural Language* (Dordrecht: D. Reidel, 1972) p. 464.

21. B. Russell, "Lectures on the Philosophy of Logical Atomism," *Logic and Knowledge* (New York: Macmillan, 1956), p. 239.

22. J. M. Dunn, "A Truth Value Semantics for Modal Systems" in *Truth, Syntax and Modality* ed. H. Leblanc, (Amsterdam: North Holland, 1973); and H. Leblanc, *Truth-Value Semantics* (Amsterdam: North Holland, 1976).

23. R. Barcan-Marcus, "Dispensing with Possibilia" *Proceedings and Addresses of the American Philosophical Association* Vol. XLIX, 1975-1976, pp. 22-38.

24. S. Kripke, "Is There a Problem about Substitutional Quantification?" in *Truth and Meaning* ed. G. Evans and J. McDowell (Oxford: Clarendon Press, 1976), p. 375.

25. Ibid., pp. 405-406.

26. H. Leblanc, "Truth Value Semantics for a Logic of Existence," *Notre Dame Journal of Formal Logic*, 12 (April, 1971), p. 154.

27. J. Hintikka, *Models for Modalities* (New York: Humanities Press, 1969), pp. 57-60.

28. H. Hiz, "The Aletheic Semantic Theory," *The Philosophical Forum*, 1 (1969), 438-451.

29. H. Leblanc, *Truth-Value Semantics* (Amsterdam: North Holland, 1976).

30. R. Martin, *Truth and Denotation* (London: Routledge and Kegan Paul, 1958), chaps. viii, ix, and x section d.

31. This argument is essentially a restatement of the one given by Andrjez Mostowski in his "On the Rules of Proof in the Pure Functional Calculus of the First Order," *Journal of Symbolic Logic*, 16 (June, 1951), p. 107.

Chapter Four

Is Existence a Predicate?

'Being' is obviously not a real predicate.

I. Kant[1]

"Existence is not a predicate" has become increasingly fashionable in philosophical circles ever since Kant. After Frege and Russell the slogan gained an even wider acceptance. However, it is not true that this tendency began with Kant, nor is it quite clear just what this claim amounts to. Kant himself said that existence is not a real predicate. Fregean and post-Fregean assertions vary from the idea that 'exists' is meaningless, to the notion that it is tautologous-analytic, redundant, universal, higher level, semantic and so on.

After Frege, we find myriad remarks on the subject:

> To be sure, it has been known for a long time that existence is not a property (cf. Kant's refutation of the ontological proof for the existence of God). But it was not until the advent of modern logic that full consistency on this point was reached.[2]

> Since Kant, we have most of us, paid lip service to the doctrine that 'existence is not a quality' and so we have rejected the pseudo-implication of the ontological argument; . . . But until fairly recently it was not noticed that . . . in 'God exists' 'exists' is not a predicate (save in grammar).[3]

> [A]s Kant pointed out, existence is not an attribute. For when we ascribe an attribute to a thing, we covertly assert that it exists: so

86

that if existence were itself an attribute, it would follow that all positive existential propositions were tautologies, and all negative existential propositions self-contradictory; and this is not the case.[4]

Russell undertook to resolve the anomalies of existence by admitting the word 'exists' only in connection with descriptions, and in explaining the whole context '$(\imath x)(...x...)$ exists' as short for '$(\exists y)(x)$ $(x = y. \equiv ...x...)$' .... This course supplies a strict technical meaning for Kant's vague declaration the 'exists' is not a predicate; namely, 'exists' is not grammatically combinable with a variable to form a matrix 'y exists'.[5]

Such statements are frequently incorporated into an analysis of existence in terms of quantification. It is appropriate at this point to give a critical history of this tradition. One of my goals here is to discover what, if anything, is common to Kant's position and more recent ones on existence and quantification. I shall explicate Kant to make clear exactly what he did say about existence. Next we shall critically evaluate his position as well as those of his twentieth century commentators. In the nineteenth century Brentano adapted Kant's view to his own new analysis of the relation of general sentences and existence sentences. We will note how this differs from earlier approaches such as those of Mill, Kant and their predecessors. With this as our background we will turn to evaluate the various arguments and claims about existence not being a predicate (especially those touched on in the first chapter). In doing this we will be in a better position to see what Kant and more recent writers have in common.

A. Kant: Existence Is Not a Real Predicate

Many discussions as to whether existence is not a predicate appear in connection with the ontological argument. The purport of most of these is to show that if 'exists' is not a predicate then a crucial premise in Anselm's argument is suspect. Perhaps this is why so many who read Kant on the subject of existence have confined themselves to the material found in the section of Kant's first *Critique* entitled: "The Impossibility of an Ontological Proof of the Existence of God".[6] I contend that this section cannot be understood without examining earlier passages of the *Critique*. At least one interpreter who did not take account of these earlier sections constructed an apparent contradiction in Kant's account.[7] I shall discuss the treatment of existence in Kant's Transcendental Analytic, especially Book II, Chapter III, section 4 ("The Postulates of Empirical Thought in General") to shed light on his discussion on the ontological argument.

I. Existence in the Transcendental Analytic

We must begin as far back as the table of judgments (B 100—for convenience' sake the references shall, where appropriate, indicate the pagination of the *Critique of Pure Reason*) to get an idea of the source of Kant's doctrine of existence. Throughout the *Critique of Pure Reason*, existence is treated with possibility and necessity as a modality. At first sight this is a quite distinctive view and constitutes an immediate difference between Kant's and other later accounts. 'Modality' for Kant, especially in his transcendental logic, has a peculiar meaning and should not be hastily identified with any contemporary account of it.

> The *modality* of judgments is a quite peculiar function. Its distinguishing characteristic is that it contributes nothing to the content of the judgment (for, besides quantity, quality, and relation, there is nothing that constitutes the content of a judgment), but concerns only the value of the copula in relation to thought in general. Problematic judgments are those in which affirmation or negation is taken as merely possible (optional). In assertoric judgments affirmation or negation is viewed as real (true), and in apodeictic judgment as necessary.... Since everything is thus incorporated in the understanding step by step—inasmuch as we first judge something problematically, then maintain its truth assertorically, and finally affirm it as inseparably united with the understanding, that is, as necessary and apodeictic—we are justified in regarding these three functions of modality as so many moments of thought.[8]

In the opening of this paragraph we find the first appearance of the doctrine that modality contributes nothing to the content of a judgment. Existence as a modal notion likewise adds nothing to the concept of an object. As we shall see, Kant will make this claim a number of times. It is important to bear in mind that for Kant this is a peculiarity of all the modal notions, and not just of existence alone.

In the sections on the categories, Kant's table of judgments is derived from general logic, while the table of categories is derived from transcendental logic. The difference between these two notions of logic is extremely important for a proper understanding of Kant's doctrine of modality as well as his later treatment of the ontological argument. General logic is what we today call formal logic. In Kant's words, general logic abstracts from all questions of the content of a judgment. Transcendental logic, Kant's unique contribution, is a decidedly epistemological endeavor which is primarily concerned with the *a priori* conditions for knowledge and which to this extent at least does not abstract from all questions of content. Existence is mentioned for the first time in the table of categories (B 106). The category or pure concept of existence is the transcendental counter-part of what is known in general logic as the assertoric mode of judging and accords with the policy of constructing the categories according to Kant's table of judgments.

To emphasize the epistemological-transcendental character of Kant's categories of modality, let us look at some passages on the schemata for the categories (B 184). In reading these we must bear in mind the Kantian doctrine that time is a form of intuition.

The schema of possibility is the agreement of the synthesis of different representations with the conditions of time in general. Opposites, for instance cannot exist in the same thing at the same time, but only the one after the other. The schema is therefore the determination of the representation of a thing at some time or other.

The schema of actuality is existence in some determinate time.

The schema of necessity is existence of an object at all times.

... Finally the schema of modality and of its categories is time itself as the correlate of the determination whether and how an object belongs to time.[9]

In the section "The Postulates of Empirical Thought in General" (B 266-294) we arrive at the fullest account of the categories falling under modality. We shall explicate this material in six parts. This order parallels the *Critique*.

1. *Brief comments on the categories of modality in general (B 266)*

In his brief opening statement there is an important characterization of existence. This is of course found alongside parallel characterizations of possibility and necessity.

1.   That which agrees with the formal conditions of experience, that is, with the conditions of intuition and of concepts, is *possible*.

2.   That which is bound up with the material conditions of experience, that is with sensation, is *actual*.

3.   That which in its connection with the actual is determined in accordance with universal conditions of experience, is (that is, exists as) *necessary*.[10]

We again find the familiar statement about modal concepts, though this time it is elaborated upon somewhat.

The categories of modality have the peculiarity that, in determining an object, they do not in the least enlarge the concept to which they are attached as predicates. They only express the relation of the concept to the faculty of knowledge. Even when the concept of a thing is complete, I can still enquire whether this object is merely possible or is also actual, or if actual, whether it is not also necessary. No additional determinations are thereby thought in the object itself; the question is only how the object, together with all its determinations, is related to understanding and its empirical employment, to empirical judgment, and to reason in its application to experience.[11]

2. Possibility (A 220) By 'possibility' Kant usually means more than merely logical possibility i.e., possibility from the standpoint of general-formal logic. To mark what he has in mind he employs the term 'real possibility' for the appropriate transcendental notion and to illustrate this difference he constructs an interesting example of something which is a mere logical possibility but is not a real or transcendental possibility.

The postulate of the *possibility* of things requires that the concept of the things should agree with the formal conditions of an experience in general. . . . [T]here is no contradiction in the concept of a figure which is enclosed within two straight lines, since the concepts of two straight lines and of their coming together contain no negation of a figure. The impossibility arises not from the concept in itself, but in connection with its construction in space, that is, from the conditions of space and of its determination.[12]

Though in the section on necessity he makes a similar contrast between real and logical necessity, it is only in connection with the ontological argument that we find the contrast between 'is' as a logical notion and 'is' as a transcendental notion.

3. *Actuality-Existence (A 225)* It is important to recognize that Kant tends to use 'actual' and 'real' and 'exists' interchangeably.[13] This, one of his most extended treatments of existence, opens by emphasizing that

> The postulate bearing on the knowledge of things as *actual* does not, indeed, demand immediate *perception* . . . of the object whose existence is to be known. What we do, however, require is the connection of the object with some actual perception, in accordance with the analogies of experience, which define all real connection in an experience in general.[14]

This clarification is necessary so that Kant can distinguish his own treatment of existence from that of those he calls "idealists." An "idealist" he has in mind is Berkeley. These idealists confine knowledge of existence to what is immediately experienced, for example, to be is to be perceived. Kant, however, extends knowledge of existence to that which is merely bound up with experience. By 'bound up with experience' he means that which is either perceived or connected in a lawlike or causal way with experience. In this respect he resembles Hume, who also held that there is mediate knowledge of existence and that it depends on the cause-and-effect relation.

However, Kant is not content with merely pointing out the difference between his view of existence and that of the so-called idealists. He adds a special subsection, "Refutation of Idealism," to his account of existence. For Kant, what exists is a representation, but he must argue against those idealists-sceptics who declare "the existence of objects in space outside us either to be merely doubtful and indemonstrable or to be false and impossible."[15] As Kant understood it, the refutation is that we can account for the existence of states of consciousness only as caused by objects in space and time.

4. *Necessity (A 227)* Kant begins this section with the distinction made between formal-logical and material-real necessity. His treatment

appears to be in the tradition of Hume, although he never mentions him:

> the necessity of existence can never be known from concepts, but always only from connection with that which is perceived, in accordance with universal laws of experience. Now there is no existence that can be known as necessary under the condition of other given appearances, save the existence of effects from given causes, in accordance with laws of causality.[16]

Like Hume, Kant is here saying that judgments of real necessity are not *a priori*, i.e., purely conceptual, but depend on experience and in particular on the cause-and-effect relation.

5. *The coextensiveness of the fields of the possible, the actual and the necessary (A 230)* One might speculate here as to whether Kant is defending the view that there is only one domain of individuals to be appealed to when making modal distinctions. Leibniz was of this opinion as are some contemporary modal logicians.[17] In any case Kant's point is one of transcendental logic, that there is only one series of appearances, i.e., that the possible, the necessary, and the actual belong to the same series.

Of greater interest to our purposes are some negative comments on a presumably Leibnizian definition of existence as a species of the possible.

> Everything actual is possible; from this proposition there naturally follows, in accordance with the logical rules of conversion, the merely particular proposition, that some possible is actual; and this would seem to mean that much is possible which is not actual. It does indeed seem as if we were justified in extending the number of possible things beyond that of the actual. But this (alleged) process of adding to the possible I refuse to allow. For that which would have to be added to the possible, would be impossible. What can be added is only a relation to my understanding, namely, that in addition to agreement with the formal conditions of experience there should be connection with some perception.[18]

Kant begins by rejecting a definition of existence by genus and differentia, wherein the possible serves as the genus. He ends by proposing a characterization of existence using this very same genus. However, the difference—what is added to the possible—is now only "a relation to my understanding," a merely subjective relation.

6. *Closing remarks (A 233)* Here Kant makes his most useful commentaries on what existence and the other modal notions do and do not add to concepts.

> The principles of modality are not, however, objectively synthetic. For the predicates of possibility, actuality, and necessity do not in the least enlarge the concept of which they are affirmed, adding something to the representation of the object. But since they are none the less synthetic, they are so subjectively only, that is, they add to the concept of a thing (of something real), of which otherwise they say nothing, the cognitive faculty from which it springs and in which it has its seat. Thus if it is in connection only with the formal conditions of experience, and so merely in the understanding, its object is called possible. If it stands in connection with perception, that is, with sensations as material supplied by the senses, and through perception is determined by means of the understanding, the object is actual. If it is determined through the connection of perception according to concepts, the object is entitled necessary. The principles of modality thus predicate of a concept nothing but the action of the faculty of knowledge through which it is generated.[19]

My reason for quoting from this section "The Postulates of Empirical Thought in General" at such length is so that I can use these remarks to interpret the better known sections on existence and the ontological argument. First, however, I offer some conjectures on the origin of Kant's treatment of existence in transcendental logic. It has long been acceptable practice to regard Kant as the heir to two traditions: the rationalist (Leibniz-Wolff) and the empiricist (Berkeley-Hume). As we saw, Kant rejected the Leibnizian definition of existence in terms of possibility only to go on to suggest the same sort of definition with his own differentia. He has continued to think of existence as a modality. Though Berkeley and Hume do not categorize existence as a modal concept, Kant's views are nonetheless closely related to theirs. Kant's remark that "that which is bound up with the material conditions of experience, that is, with sensation, is actual" must be contrasted with Berkeley's "to be is to be perceived". The similarity is that existence is connected with experience; the difference is that Kant says that what exists need not be perceived but merely bound up with experience. By 'bound up', Kant seems to have meant that it has some lawlike-causal connection with experience. In this respect he is following Hume, who said that questions of existence depend on the cause-and-effect relation. Kant himself seems to have been aware of the similarity of his view and Berkeley's since he included a section on the refutation of idealism as part of his treatment of existence. The refutation certainly indicates a difference between his and Berkeley's views and thus obviates any charge of his being an idealist in some extreme sense.

Another influential factor is Berkeley's insistence that we have neither abstract nor particular ideas of existence. Hume also said that "the idea of existence is not derived from any particular impression" and therefore, "that idea, when conjoined with the idea of any object, makes no addition to it."[20] Although these remarks account for Kant's view that existence adds nothing objective to the concept of an object, two questions remain: (1) Why did Kant say the same thing for necessity and possibility? (2) Why did he say that what is added is something subjective rather than that nothing whatsoever was added at all? (That nothing was added at all is consistent with the treatment of existence in formal logic in terms of the assertoric mode of the copula.) The answer to these questions is no doubt likely to be found in Hume's influence. Hume agreed with Berkeley that we have no objective impression-idea of existence; he then extended this analysis to necessity. The idea of necessary connection is not objective, i.e., it is not a copy of some external impression. Kant probably saw this similarity between the idea of necessity and that of existence, and then said the same for the third modal notion of possibility. His repeated declaration that modal notions, in particular that of existence, add nothing objective to a concept but only something subjective, appears to be an extension of one of Hume's ideas. Hume said that the idea of necessary connection has its origin in an internal impression, i.e., the gentle force present in anticipating that a like effect will follow a like cause. Kant's insistence that judgments of existence are synthetic but that what is added to the subject concept is only subjective is perhaps best viewed against this Humean background.

II. The Impossibility of an Ontological Proof of the Existence of God

Here Kant takes up the subject of an absolutely necessary being. His material is divided into several sections.

1. General considerations about existence and subject-predicate judgments. (A 592-596).
2. A purported exception to the above general considerations viz. the case of one unique concept "in reference to which the not-being or rejection of its object is itself contradictory". (A 596-597)
3. Kant's solution to the problem posed by this purported exception.
   A. Some idle and fruitless disputations rooted in the confusion of logical and real predicates. (A 597-598)
   B. Being is not a real predicate (A 598 to end).

1. General considerations about existence and subject-predicate judgments In his opening paragraphs, Kant considers the verbal definition of a necessary being as something the non-existence of which is impossible. In effect Kant is remarking on the sentence 'A necessary being is one whose non-existence is impossible' from the standpoint of general logic. The general or traditional formal logic he inherited from Wolff and Leibniz taught that affirmative propositions have existential import. If 'A triangle has three angles' were true, this would entail that there are triangles. Kant points out that if there were triangles, then they necessarily would have three angles; but one can nonetheless deny the existence of triangles. That is to say, an affirmative proposition can be false either because the referent of its subject term does not exist or, even if the referent does exist, if its predicate does not truly apply to its subject. Similarly, our sentence about a necessary being can be rejected in the sense that we deny that its existence condition has been met. This provides an explanation of the following passage within the context of the formal logic of the time.

If, in an identical proposition, I reject the predicate while retaining the subject, contradiction results; and I therefore say that the former belongs necessarily to the latter. But if we reject subject and

predicate alike, there is no contradiction; for nothing is then left
that can be contradicted. To posit a triangle, and yet to reject its
three angles, is self-contradictory; but there is no contradiction in
rejecting the triangle together with its three angles. The same holds
true of the concept of an absolutely necessary being. If its existence
is rejected, we reject the thing with all its predicates; and no question of contradiction can arise.[21]

2. A purported exception to the general considerations After these
considerations from general logic, Kant raises a purported exception to
them, viz. the unique case of a concept whose object must exist. This
is the concept of a being which contains all reality; since reality includes existence, existence is part of this concept (A 596-597). (It is as
though one were to argue that though 'if *x* is perfect being, then *x* exists' is not a logical truth in the strictly formal sense, it is so in some
extended sense.) Kant immediately voices his opinion that a contradiction is embedded in a concept which includes existence. He takes the
trouble, however, to entertain this concept in order to show how it
leads to "idle and fruitless disputations."[22] He considers an existential
judgment, e.g., 'This exists', and then constructs a dilemma on the basis
of the alternatives of whether the judgment is analytic or synthetic. (It
is important to note in passing that most of Kant's examples of existence statements are singular, e.g., 'This exists', or 'God exists'.) If this
sentence is analytic then 'exists' would add nothing to the concept
involved. But this horn of the dilemma yields two possibilities. On one
hand, the concept correlative to the word 'this', " . . . the thought in
us . . ." is the thing itself. On the other hand, we have presupposed an
existent and the judgment is a mere tautology. It is not easy to see how
Kant has made his case for this horn of the dilemma, but he himself
seems to be satisfied with his argument that the supposition that 'This
exists' is analytic leads to difficulties. The alternative that 'This exists'
is synthetic (which is Kant's own view), contradicts the hypothesis
that the concept of existence is already included in the concept in
question. However, since Kant himself dismisses these speculations as
idle and fruitless, we will not be unduly concerned at our failure to
understand this material.

3. Kant's solution to the problem The confusion which led to the
acceptance of this troublesome concept came from treating existence

as a real or determining predicate simply because it sometimes appears as a logical-grammatical predicate. A determining predicate is one which enlarges the concept of the subject. To put the matter differently, the confusion rests on failing to distinguish the different roles of 'being' in general and transcendental logic. With regard to general logic Kant says two things: (1) anything we please can serve as a logical predicate, and (2) 'being' logically plays the role of the copula in the subject-predicate judgment. Since anything can serve as a predicate in general logic, when 'being' plays this role it does not afford a clue to its significance in transcendental logic. However, when 'being' plays the role of a copula in general logic, it is not a predicate on a par with other predicates and so does furnish a clue to its role in transcendental logic. 'Being' as the copula indicates the mode in which the predicate is in a subject. It is a *de re* modality (a mode of the copula, not the sentence) and on a par with the other modal pseudo-predicates 'possibility' and 'necessity'. This is the basis in general logic for the treatment of the category of modality in transcendental logic. Thus Kant's rejection of the ontological argument rests on his view that existence *qua* modal notion is not part of any concept.

The claim Kant makes that existence in general-formal logic is connected with the assertoric mode of the copula has a long history. This view is connected with the traditional conception of existential import mentioned above; that affirmative propositions have existential import, i.e., from *a* is a *b* it follows that *a* is and that *b* is. One might interpret Kant as saying that 'being' should be analyzed in terms of the copula. To draw a parallel with a claim made for the "existential" quantifier, Kant can be construed as claiming that 'exists' in general logic is not a genuine predicate, since existence sentences are contextually definable in terms of other sentences in which the expression 'exists' no longer appears and in which its function is now performed by a logical constant, viz., the copula. In appendix A I will take up precisely this view, for it finds a contemporary exponent in the person of Lesniewski. One of Lesniewski's systems of logic, called Ontology, defines 'exists' in just this way. The system is an extended logic of terms with a form of the copula serving as the primitive constant.

It seems impossible to understand correctly the famous passage (A 599) explaining that "being is manifestly not a real predicate" without bearing in mind the distinction between general and transcendental logic and recognizing that 'being' from Kant's standpoint is a modal notion.[23]   Having dealt with the ontological argument in terms of general logic in the opening paragraphs (A 592-596) Kant is now offering a treatment of the problem in terms of the nature of modality in transcendental logic. Thus when he turns to the sentence 'God is' we find him making the same sort of remarks we found in the "Transcendental Analytic." When we say 'God is', "we attach no new predicate to the concept of God, but only posit the subject itself with all its predicates, and indeed posit it as being an object that stands in relation to my *concept*." Again he says in explanation of the difference between the concept of a hundred real and a hundred possible thalers:

> For the object, as it actually exists, is not analytically contained in my concept, but is added to my concept (which is a determination of my state) synthetically; and yet the conceived hundred thalers are not themselves in the least increased through thus acquiring existence outside my concept.

This then is merely reiterating that 'exists' as a modal notion synthetically adds something to the concept, but only in a subjective way. Thus far Kant's treatment of existence in this famous passage (and for that matter in the entire section) has brought forth no distinctive contributions. Up till now Kant has either relied on the general logic of his time or applied the transcendental account of existence already given earlier in the *Critique of Pure Reason*. Now, however, we find what appears to be a significant new addition: an argument in two forms and two examples, all to the effect that existence does not add anything to the concept of an object. The importance of these arguments and examples cannot be over-emphasized since they are often cited as independent grounds for the claim that 'exists' is not a real predicate. We are first given the famous example that "A hundred real thalers do not contain the least coin more than a hundred possible thalers." Then in connection with this example, Kant offers a *reductio ad absurdum* argument. On the assumption that an existing x (a hundred thalers) contained more than a possible x (where the latter is supposed to be the concept of the former), the absurd consequence follows that the concept would not be adequate to the object. The argument is immediately restated in the very next paragraph (A 600) to produce the absurd result that we could never make the existential judgments we want to. If in judging that a thing exists we make an addition to the thing, then it would not be "exactly the same thing that exists, but something more than we had thought in the concept; and we could not, therefore, say that the exact object of my concept exists." Next Kant gives an illustration of the argument's point. "If we think in a thing every feature of reality except one, the missing reality is not added by my saying that this defective thing exists. On the contrary, it exists with the same defect with which I have thought it, since otherwise what exists would be something different from what I thought." If one has already accepted Kant's modal account of existence in transcendental logic, these arguments appear plausible, but they then lose their significance as providing independent evidence.

The remainder of this section highlights the peculiar difficulty involved in asserting the existence of a supreme being. For Kant a supreme being, like any other being, exists if its concept is in the right sort of epistemological relation to something outside itself. Because the concept of a supreme being is not empirical, but is one of pure thought, it has neither an *a posteriori* nor an *a priori* relation to something outside itself. Hence the existence of a supreme being is "an assumption which we can never be in a position to justify".

The reader who accepts the above interpretation of Kant, as holding that in general logic existence claims are translatable as "copula" claims in the assertoric mode, can pursue this parallel with contemporary views, and especially with Quine's. The common theme is that general logic, either that of the copula (Kant-Lesniewski) or of the quantifier (Russell-Quine), provides the vehicle for expressing existence claims. The independent and additional issue of how one adjudicates between existence claims is Kant's subject in this area of transcendental logic. Kant is unlike Berkeley, whom I interpreted as saying that decisions about ontological issues must be made solely on the basis of direct evidence, i.e., perception. Kant is in this respect a more modest empiricist for whom ontological decisions are made on bases of direct or indirect evidence, i.e., to exist is to be bound up with experience in a lawlike way. Quine has an even more liberal criterion for deciding ontological issues: his Duhemian conception of the scientific method. All three, Berkeley, Kant, and Quine, are construed as empiricists on the question of adjudicating between ontological claims. The upshot of

Kant's critique of the ontological argument is typically empiricist; the concept of a supreme being has no empirical support—the concept of such a being is not deductively or inductively bound up with any experience.

III. A Critical Evaluation of Kant and a Current Interpretation

In a much-reprinted article, Jerome Shaffer claims to have discovered a contradiction in Kant's account.[24] According to Shaffer, Kant cannot consistently maintain (1) that 'exists' is not a real predicate and (2) that existential judgments are synthetic. If 'exists' were not a real predicate, then it would not enlarge the subject concept of an existential judgment. But a judgment where the predicate concept does not enlarge the subject concept is analytic. It should be obvious to the reader how Kant would reply and how Shaffer arrived at his mistaken interpretation.[25] Kant would point out, as he did explicitly in the "Transcendental Analytic," that modal predicates add something subjective to the concept of an object, i.e., a relation to the faculty of knowledge. (This is also implicit in the last part of the section on the ontological argument where we are told that what is added to the concept is a determination of one's state.) Shaffer and many others misinterpret Kant for at least two reasons. They ignore the earlier sections of the *Critique* and go on to read the later parts out of context. A related error by Shaffer *et al.* is that they do not observe Kant's distinction between general and transcendental logic. They thus do not realize that much of what Kant says about existence is a part of transcendental logic and not, as Shaffer takes it, part of a general theory of predication. Furthermore, to whatever extent Kant does discuss issues of general logic, they must be viewed in terms of the general logic of his time.

How are we to assess Kant's doctrine? I will concentrate on two points, (1) his modal conception of existence as part of transcendental logic, and (2) a re-examination of the arguments and examples that 'being' is not a real predicate.

Let us recall some of the modal descriptions of existence from the standpoint of transcendental logic.

That which is bound up with the material conditions of experience, that is, with sensation, is *actual*.

In the *mere concept* of a thing no mark of its existence is to be found. For though it may be so complete that nothing which is required for thinking the thing with all its inner determinations is lacking to it, yet existence has nothing to do with all this, but only with the question whether such a thing be so given us that the perception of it can, if need be, precede the concept.

[W]hatever is connected with perception in accordance with empirical laws is actual, even though it is not immediately perceived.[26]

In the first place, to whatever extent Kant's treatment of existence is rooted in his transcendental methodology, it is problematic. The questionable character of that method would provide a telling criticism in itself. However, there is another line of criticism which I shall pursue, concerning a common feature of Kant and his British predecessors. Kant says that existence is an epistemological relation, i.e., x exists if and only if the concept of x is bound up with experience.[27] Though he has succeeded in distinguishing his account of existence from Berkeley's definition—to be is to be perceived—nonetheless an important similarity remains. A criticism frequently made of Berkeley's characterizations is that 'exists' cannot be correctly defined in terms of the perceivable because there is no contradiction in saying that something exists but is not perceivable. Just so, there is no contradiction in saying that something exists but is not bound up with experience. Furthermore Berkeley and Kant seem not so much to be giving a definition of existence as trying to provide a criterion for the truth of the claim that something exists. Indeed much of Kant's transcendental logic concerns itself with the conditions for knowing. This point is embodied in the suggestion I made that Kant in general logic is addressing the question of how we express existence claims (by the use of the copula), while in transcendental logic we take up the question of how we decide on existence claims (by the use of an empiricist's methodology).

Let us turn now to a closer critical look at the famous examples and arguments given in connection with the ontological argument. To begin, consider Norman Kemp Smith's rendering of Kant's famous example.

Otherwise stated the real contains no more than the merely possible. A hundred real thalers do not contain the least coin more than a hundred possible thalers.[28]

This example is really of no help since it is completely irrelevant that the purported difference between the real and possible thalers should be one of amount.

Kant uses this example to present an argument showing that existence cannot be a real predicate. The same argument is given in a slightly varied form in the next paragraph. These arguments are terribly difficult to understand, and even though they are continually referred to in the literature, very little light has been shed upon them. This is evidenced by the fact that they are almost always quoted *verbatim*. In the following restatement I restrict myself to Kant's terminology of 'concepts', 'concepts being adequate to an object', etc. Although these notions are unclear, I cannot dispense with them and still do justice to Kant's intentions.

Both arguments are of the *reductio ad absurdum* type. Let us concentrate on the first.

(1) We have an adequate concept of an object x where x exists (Note that x as existing $= x$).

Here Kant is interested in comparing concepts with objects. In an adequate concept (it is only these which are questioned here), there is nothing more in the concept than in the object, or vice versa. Let us add then an additional assumption.

(2) An adequate concept of an object contains all and only the true descriptions of that object.

We note in passing that this requirement for a concept being adequate is inordinately stringent. That is to say, we would ordinarily allow that someone had an adequate concept for an object without requiring that every aspect of that object was part of the concept.

(3) Existence changes x, so that x as existing is altered to y where $y \neq x$.

(4) The originally adequate concept of x is no longer adequate.

Kant concludes that assumption (3), that existence changes x, is responsible for the absurdity and thus claims that existence does not change x; in other words, existence is not a determining property.

Shaffer has raised the interesting criticism that if the argument were good, then it would prove too much—it would establish that there are no real (determining) predicates. This point can be made by generalizing on the original argument.

(5) We have an adequate concept of an object x where x is an F (Note that x as an $F = x$).

(6) An adequate concept of an object contains all and only the true descriptions of that object.

(7) F changes x, so that x as an F is altered to y where $y \neq x$.

(8) The originally adequate concept of x is no longer adequate.

Shaffer's case is disquieting, because though it presents a damaging criticism, it tells us nothing as to the locus of Kant's error, which derives from thinking that the absurdity follows solely from the assumption that F changes x and that one can then legitimately conclude that it is not true that F changes x. In fact the absurdity follows from holding at once that we have an adequate concept of x and that F changes x. (5) and (7) are inconsistent since in (5) we are told that x as $F = x$ whereas in (7) we are told that x as $F = y$ where $x \neq y$. In fact from (5) and (6) alone (the first two premises and presumably the very ones which Kant does not wish to question) it follows that if we have an adequate concept of x and x is an F, or in particular x exists, then either F or existence is included in that adequate concept. For example, if we have an adequate concept of Socrates then since Socrates exists, existence is part of that concept. If we had an adequate concept of Cerberus, then since Cerberus does not exist, existence is not part of that concept.

Kant offers an example to illustrate the point of his second argument. Earlier he had contrasted a hundred real and a hundred possible thalers. Here he contrasts the concept of a defective being (e.g., the concept of George Washington as not having Martha as his wife is defective since in reality she was his wife) with the concept of the defective being as existing.

> If we think in a thing every feature of reality except one, the missing reality is not added by my saying that this defective thing exists. On the contrary, it exists with the same defect with which I have thought it, since otherwise what exists would be something different from what I thought.[29]

Like the hundred thaler example this purported illustration is quite irrelevant. No one has claimed that the difference between the concept of a defective thing and the concept of a defective thing as existing is that the latter is no longer a concept of something defective.

The moral I draw from this investigation is that 'exists' can be treated as a real predicate, as adding something to an object, and that we can have an adequate concept of that object if it contains the concept of x as existing. Kant's arguments and examples were intended to show that 'exists' is not a real predicate. They do not succeed in doing this nor in making a case for his modal conception of existence.

B. Nineteenth Century Discussions of General Sentences and Existence

Between the time of Kant and that of Peirce and Frege the subject of existence and logic was by no means neglected. Brentano made an important contribution here. He seems to have been the first philosopher to claim that all general sentences are reducible to existential sentences.[30] He explicitly rejected Mill's and Kant's traditional classification of sentences. For Mill, and quite likely for the history of logic since Aristotle, non-compound sentences were classified as general, i.e., universal or particular, and singular. According to this older tradition, existence sentences could occur, without anomaly, either as universal, e.g., 'All the pieces of the puzzle exist', or particular, e.g., 'Some of the pieces of the puzzle don't exist', or as singular, e.g., 'This piece of the puzzle exists'. Brentano argued in 1874 that all sentences are merely varieties of existential ones. The argument he gives is for categorical sentences of the A, E, I, and O form, but he claims (without justifying it) that all sentences, including conditional and apparently singular ones are similarly reducible. I will deal only with the claim for the reducibility of categorical sentences. Brentano argues that all categorical sentences are synonymous with corresponding existential ones.

	Categorical	*Existential*
I - form	Some man is sick.	A sick man exists.
		There is a sick man.
E - form	No stone is living.	A living stone does not exist.
		There is no living stone.
A - form	All men are mortal.	An immortal man does not exist.
		There is no immortal man.

O - form   Some man is not learned.   A non-learned man exists.

There is a non-learned man.

This argument was mentioned in the first chapter where it was used to show that the function of 'exists' can be fulfilled by using quantifiers.

Unlike later authors, Brentano presents this argument without any of the benefits of modern logic.

It is important to stress the difference, with regard to the place of existence in logic, between authors like Mill, Kant and their predecessors, on the one hand, and Brentano on the other. For Mill, 'exists' is a non-logical predicate constant, in the current sense of that word. For Kant, 'exists' in general logic, because of the relation to the copula, differs from ordinary predicates, i.e., it is not a real predicate. Brentano was among the first to take Kant's claim about 'exists' not being a real predicate and apply it to the general-formal logic of quantifiers.

Another important concern of nineteenth-century philosophy and logic is the topic of 'existential import'. I mentioned Kant's approach to this question earlier. Though Kant was in this respect only following Leibniz, the tradition is much older.[31] It held that the quality of a proposition determines its existential import. Affirmative sentences had existential import and negative ones did not. An A- or I-form sentence with a vacuous subject term was false, while E- and O-form sentences with vacuous subjects were true. After the work of such nineteenth-century figures as Boole, De Morgan, Venn, Brentano and others the situation was quite different. According to the Boolean interpretation, quantity determines existential import. Particular sentences with vacuous subject terms are false while universal sentences of this sort are true. Now all particular sentences have existential import.

This nineteenth-century notion of existential import helped to lay the foundation for Frege and others' ready acceptance of the view that 'exists' has some special link with the particular quantifier. By the third quarter of the nineteenth-century the view that existence is entailed by particular generalizations came to be rather widely held. This modification of traditional logic was incorporated into the modern logic of Frege, Peirce, Russell and their followers.

It is interesting to conjecture as to the justification of the view that quality and not quantity conveys existential import. Benson Mates has cited the scholastic principle *"Nihile nullae proprietates sunt"* that he discovered in Leibniz's work.[32] This can be translated to say that nothing has no properties, and might be expressed quantificationally as

$$(x)(\sim(\exists y)\, y = x \supset (\phi)\sim\phi x)$$

On the existential reading, it would say that what does not exist has no properties. Mates says that to accept this principle is to regard singular sentences with vacuous subjects as false. For instance, assuming that Pegasus does not exist, i.e., $\sim(\exists y)(y = \text{Pegasus})$, it follows that any simple sentence in which something is predicated of Pegasus would be false.

I would like (1) to suggest that this principle might justify the older tradition of existential import and (2) to point out how current authors, especially Russell and Quine, might accept this principle for singular sentences but not for general ones. As early as Albert of Saxony (1353) we find truth conditions given for general sentences; universal sentences are treated as conjunctions and particular ones as disjunctions. For instance:

All men are mortal	This man is mortal & That man is mortal & etc.
Some men are not mortal	This man is not mortal **v** That man is not mortal **v** etc.

Now, if vacuous singular sentences are false, then the conjunctions and disjunctions which express the truth conditions for affirmative sentences would never be true. A negative singular sentence with a vacuous subject would always be true, and so the expansions for E- and O-form sentences would under these circumstances be true.

The contemporary Boolean expansion of such sentences is quite different:

All men are mortal	If this is a man then this is mortal & if that is a man then that is mortal & etc.
Some men are not mortal	This is a man and this is not mortal **v** That is a man and that is not mortal **v** etc.

Here the constituents of the conjunctions and disjunctions are not singular sentences but conditionals (for the A- and E-form) and conjunctions (for the I- and O-forms). A number of current writers treat singular sentences as always false when the subject is vacuous. This is indeed true for those who hold Russell's theory of descriptions. But given the modern expansions for general sentences, one could hold the Boolean view for them while maintaining the scholastic principle for singular sentences.

One virtue of the older tradition on expansions of categorical sentences is that it allows one to account for the truth of some O-form existence claims. Consider the plausible candidate: 'Some of the horses referred to by the ancients do not exist'. An expansion of this sentence would exhibit the following form:

~(This horse mentioned by the ancients, viz. Pegasus, does exist)

or

~(That horse mentioned by the ancients does exist)

or

etc., etc.

Since the first disjunct is true, the disjunction is too, and thus the O-form sentence is true as well. The first disjunct is true, because the sentence 'This horse (Pegasus) mentioned by the ancients exists' is false, and hence its negation, the first disjunct, is true.

In summary then we note that there were two strands in the nineteenth century which probably led to the confident acceptance of the view that 'exists' and '∃x' are closely linked. The first is the Kantian doctrine that 'exists' is not a predicate, which was conveniently if inaccurately taken as evidence that it is a quantifier. The second is the Boolean conception of the existential import of categorical propositions, according to which existence is conveyed by particular sentences.

C. Is 'Exists' Meaningless, a Tautological Predicate, A Universal One, A Higher Order One, A Disguised Quantifier, etc?

I shall now critically review the arguments presented in Chapter One concerning existence. The statement that 'exists' is not a predicate is simplistic. Several quite distinct claims have been made about existence sentences. Some of these concern only singular existence sentences, while others apply to general existence sentences as well; and a variety of different theses have been argued. They may be arranged into the following sorts:

I.   'Exists' is a meaningless predicate, i.e., certain existence sentences are meaningless.

II.   'Exists' is a tautological or analytic predicate.

III. 'Exists' is a universal predicate.

IV.   'Exists' is a higher order predicate.

V.   'Exists' is not a descriptive property.

VI.   'Exists' is a disguised quantifier.

For each of these I will indicate both how it differs from and what it has in common with Kant's doctrine.

I wish both to discredit the spurious claims made about existence which rest on the existential reading of the quantifier and to argue against the existential reading itself. The claims of group VI are, in effect, equivalent to the claim that this reading is correct. I have presented above some viable non-existential readings of the quantifiers. In the remainder of this monograph I will argue that these readings are superior to the existential reading.

In doing so I argue for the autonomy of existence and of the logic of quantification, i.e., their mutual independence. While I leave open the possibility that Kant, Lesniewski, and the ancient tradition they continue may be right in connecting existence and the logic of the copula, my thesis is primarily that there are good reasons for regarding 'exists' as an ordinary, if philosophically interesting, non-logical predicate constant. Such a thesis frees logic of some of its more obvious connections with ontology and gives new meaning to Ernest Nagel's phrases "logic without ontology" and "logic without metaphysics."

I. 'Exists' Is a Meaningless Predicate

The view that certain existence sentences are meaningless has been stated most strongly with regard to singular existence sentences. The reasons for this claim can be classified into two groups; those utilizing the principle of significant negation and those involving some sort of type or category considerations.

The reasoning in which the principle of significant negation plays a crucial role concerns the claim that a sentence like 'This exists' cannot be meaningful, because its negation, 'It is not true that this exists' is meaningless. The argument, however, depends not only upon the principle of significant negation, but also upon a certain account of the meaning of singular terms, in which the meaning of a singular term is the object it names. Hence, 'this' is meaningful if and only if 'this' names something. Since 'this' in the negative context does not name any existing thing, it would seem to follow that the negation is meaningless.

There are at least two reasons for rejecting this argument. The first and better known one is that the "Fido-Fido" theory of meaning upon which it depends has been roundly criticized. Whatever meaning may be, it is not, even in the case of names, the object named. A second criticism is that even on the "Fido-Fido" theory, it does not necessarily follow that the negative sentence does not name something and is therefore meaningless. I have encouraged a certain skepticism above about equating 'there exists' and 'there is something'. In particular, it does not follow from the fact that 'this' does not name some existing thing, that 'this' does not name something *simpliciter*; were something (albeit non-existent) named, the negative sentence in question would be meaningful. Another way of putting this point is to note that a "Fido-Fido" theorist might hold to the semantic principle that to refer to x, x must be something; but this is not the same as the stronger semantic principle that to refer to x, x must exist. It is easy to obscure this point if one assumes at the outset that 'exists', 'some', and '$\exists x$' are interchangeable.

The second argument occurs in Frege and Russell. It claims that 'exists' as a disguised existential quantifier is not a predicate of individuals. For Frege, 'exists' is a predicate of concepts or functions and so is a higher level concept-function. For Russell, it is a higher-level predicate of propositional functions.

Frege himself offered two arguments to show why such sentences are improper. The first purports to show that singular existence claims rest on the fallacy of division. To make the strongest case, consider again Russell's examples.

Men exist.
<u>Socrates is a man.</u>
Socrates exists.

The thing that there are in the world exist.
<u>This is a thing in the world.</u>
This exists.

Men are numerous.

Socrates is a man.
_____
Socrates is numerous.

What are we to make of this point? If we candidly consider the argument independent of any prior considerations, the supposed analogy is far from obvious; there is certainly no *prima facie* case for it. Presenting these arguments outside the context of Frege and Russell's special assumptions helps us to gain a proper perspective. It is one of the ironies of philosophy that so many contemporary thinkers have convinced themselves that singular existence sentences are somehow linguistically irregular. For Aristotle the existence of individuals was the primary sense of 'existence'. Traditionally, general and singular existence sentences were not regarded as being in any way deviant. Indeed, we mentioned that most of Kant's examples of judgments of existence were singular.

Frege's second argument is based on the principle of the inverse variation of extension and intension. If 'exists' were a predicate of individuals, its extension would be universal. But then by the principle above, it would have no intension. Two points can be made here. The argument assumes that if 'exists' were a predicate of individuals, it would be universal. It is not enough merely to assume this. Some sort of reason must be given. I shall, when considering the view that 'exists' is a universal predicate, actually present evidence to the contrary. In addition, the argument is questionable in so far as it rests on the questionable principle of inverse variation of extension and intension.

Perhaps the most compelling reason that Russell rejected singular existence sentences as meaningless has its roots in his theory of types. Russell subscribed to the ramified theory and not just the simple theory of types.[33] According to the simple, but not the ramified, theory propositional functions (or predicates) which have the same argument are of the same type. Thus the propositional functions 'x is an emperor' and 'x has all the properties necessary for being a great general' are of the same type according to the simple theory, since they both apply to the individual Napoleon. However, they are of different orders according to the ramified theory: 'x has all the properties necessary for a great general' is of a higher order because in this function we find a quantifier and a variable, i.e., 'all the properties', of the same type as 'is an emperor'. That is, in the second propositional function we are quantifying over (or talking about) predicates like 'is an emperor'. How does this apply to Russell's idea of singular existence? For Russell 'E! $\imath x \phi x$', i.e., '$\imath x \phi x$ exists', is meaningful but 'E!a' is not. In other words existence can be predicated of objects introduced by definite descriptions but not by proper names. On the ramified theory, '$\imath x \phi x$' and 'a' are not of the same order. The definition of the definite description '$\imath x \phi x$' contains quantifiers and variables for expressions of the same type as 'a'. Hence if 'E!' significantly applies to '$\imath x \phi x$', then it cannot do so for 'a'.

We have at least three alternatives to Russell's view. The first is to give up the ramified theory of types and make do with the simple. The second one is to dispense with type theory altogether. The third is to adopt a non-existential reading of the quantifier and abandon the attempt to give a purely logical definition of 'exists'.

II. 'Exists' Is a Tautological Or Analytic Predicate

The view that singular existence sentences are analytic or tautological has been argued by using a variant of the principle of significant negation, which I have called the principle of significant affirmation. Accordingly, if 'this' is meaningful if and only if 'this' names something, then 'this exists', if meaningful, is analytically true and its negation is analytically false. Such an argument is open to the same objections mentioned in connection with the use of the principle of significant negation.

While Moore claimed to show that 'exists' in general existence sentences is meaningless, I have revised this to a weaker claim that it is merely analytic or redundant. The first of Moore's arguments extended the use of the principle of significant negation to general sentences. Recall that he contrasted genuine A- and O-form sentences about growling tame tigers with "pseudo" ones about existing tame tigers.

A   All tame tigers growl.

O   Some tame tigers don't growl.

Since the O is a significant negation of the A, the A is significant too.

Pseudo A   All tame tigers exist.

Pseudo O   Some tame tigers don't exist.

Pseudo O'   There are tame tigers which don't exist.

We assume that the Pseudo O and the Pseudo O' are equivalent. Moore says that since the Pseudo O' is meaningless (or at the very least contradictory) then so is the Pseudo O. Consequently the Pseudo A is meaningless or a tautology.

Recently, several authors have pointed out that O-form existence sentences, e.g. the above "Pseudo" O, are not meaningless or contradictory and that A-form sentences are not analytic.[34] These writers are simply exhibiting a pre-Brentano view of the matter. Why then did Moore hold such a position? The most plausible explanation is that he regarded the expressions 'some', 'there is', and 'there exists' as equivalent. He probably did this because they all served as readings for the same technical expression '$\exists x$'. When I contrasted the semantic frameworks for 'some' and 'there is' with the one for 'there exists', I asserted that the frameworks for the non-existential readings provided a theoretical justification for our common-sense view that O-form existence sentences need not be analytically false. Moore's argument has force only if one adopts the existential reading. We then have a clear-cut contradiction: there exist tame tigers which don't exist. But of course the argument now depends on showing that one *must* adopt only the existential reading. Ironically this argument will later furnish a reason for not adopting that reading.

Moore's second argument is that we must dispense with using 'exists' as a predicate. The crucial part of this argument is a string of three sentences.

(9)   Some tame tigers exist.

(10) For at least one value of x, 'x is a tame tiger and x exists' is true. 10 follows from 9 and from 10 at least one sentence such as 11 follows.

(11) This is a tame tiger and this exists.

By relying on the assumption that singular existence sentences such as the second conjunct of (11) are meaningless or redundant, Moore concludes that 'exists' in (9) is meaningless or redundant as well. This argument fails because it presupposes questionable arguments for the meaninglessness or analyticity of singular existence sentences.

The claim that 'exists' is tautological was also sponsored by Reichenbach and Salmon and Nakhnikian. Reichenbach assumed that '∃x' is to be read as 'there exists' and then proposed treating singular existence so that 'y exists' is a reading for '(∃x)(x = y)'. He pointed out that sentences such as '(∃x)(x = y)' are analytic in the sense that they are consequences of the logical truth $(x)(x = x)$. Along somewhat the same lines, Salmon and Nakhnikian concluded that existence is necessarily universal and redundant, that from the above logical truth (x) $(x = x)$ one could derive both that everything exists, i.e., $(x)(∃y)(x = y)$, and that of necessity everything exists, i.e., $\Box(x)(∃y)(x = y)$. With the advent of free logic neither Reichenbach's nor Salmon and Nakhnikian's derivations are permissible. In one sense of a free logic we must allow for empty individual constants. So with Hintikka's rules, we cannot infer $(∃x)(x = y)$ from $(x)(x = x)$ alone. The former sentence is not analytic. In the second sense of a free logic (prohibiting so-called existence theorems), neither $(x)(∃y)(x = y)$ nor $\Box(x)(∃y)(x = y)$ are logical consequents.

Before leaving our consideration of whether existence sentences are analytic let us note that the notion actually conflicts with Kant's view that 'exists' is not a real predicate. For him, judgments of existence are synthetic. If in saying that 'exists' is a redundant predicate, we mean that the judgment is analytic or that the predicate is dispensable in the Frege-Russell tradition, then Kant would disagree.

III. 'Exists' Is a Universal Predicate

To show that 'exists' is a universal predicate, we would have to prove '$(x)(x$ exists)'. If 'exists' is treated as a non-logical notion, i.e. as an ordinary though philosophically interesting predicate constant, then the case is yet to be made. If on the other hand it is treated as a logical notion, i.e. treated in terms of quantification as the Frege-Russell tradition does or in terms of the copula as Kant and among others Lesniewski did, then we must recall that '$(x)(x$ exists)' is not provable in free logics.[35]

Some writers treat 'exists' as a primitive and seem to give it the status of a logical notion, but they do so without providing a justification.[36]   Why indeed should existence be a logical notion on a par with negation, disjunction, universal quantification, etc.?

IV. 'Exists' Is a Higher Order Predicate

When *Principia* was written, no distinction was made between a purely logical theory of types and a semantical theory of types. Russell, for instance, thought he was presenting at the same time both a solution to his paradox of classes and a solution to paradoxes like Eubulides' paradox of the liar.   It was not until some time later that Ramsey distinguished the purely logical from the semantical paradoxes. I submit that the claim that existence *qua* the particular quantifier is a higher level property was accepted and continues to be so not because of considerations arising from the theory of logical types proper, but rather from semantical considerations.   Recall that Russell says that 'Lions exist' i.e., '$(\exists x)(x$ is a lion)' is really a predication about the propositional function 'x is a lion', that the propositional function is sometimes true or that the class of lions has a member.   Frege also held that '$\exists x$' expressed a higher level property, although he apparently did not enunciate a theory of logical types.   Furthermore, consider the thoughts of some contemporary figures:

> the *force* of the word 'exists' is not to add some new description, ... It is not descriptive because it is semantical.   To say that 'x exists' is to say that 'x' is instantiated.[37]

and by the same author

> For 'a exists' is a *semantically* not descriptively augmented name. And the deep reason Berkeley could not form an idea of existence is because the word 'exists' is semantical, having application finally to the connection between descriptive words and what confers semantical values on them, without being a descriptive word in its own right.

> ... an existential statement is to be analyzed as an assertion that some second-order entity has a certain logical feature; for example, that a class does (not) have members, that a property is (not) instantiated, or that a propositional function is sometimes (or never) true when transformed into a proposition by instantiation of its individual variable.[38]

What do these contemporary authors have in common with the view that existence is a higher level property? Frege says that to say something exists is to say that a concept is not null. 'Lions exist' says, for Russell, something about the propositional function 'x is a lion'. Danto says that 'exists' is a semantical word and Welker summarizes each of these views. I shall argue that all of these statements about existence or existentially construed quantifications involve a confusion similar to one noted in Chapter Two. The confusion consists of (a) conflating an equivalence claim with an identity claim (or confusing a contextual definition of an incomplete symbol with an explicit definition) and (b) being misled by the semantic ascent frequently a part of discussions of logical constants.

The equivalence of the following three sentences is essential to the existential reading of the particular quantifier. I will assume the equivalence for the moment to show that even assuming it, the additional claim that existence is instantiation or some higher level (or semantical) property does not follow.

Existential sentence: Horses exist.

Quantificational Counterpart: $(\exists x)(x$ is a horse)

Informal truth conditions for the existential reading: 'Horse' designates (or the associated propositional function is instantiated, or the concept of a horse applies).

The first, the existential sentence, is contextually defined in terms of the second (the existential quantification), that is, every sentence containing 'exists' can be translated into another sentence, a particular generalization, in which the word 'exists' no longer appears. The third sentence is intended as a rough statement of the truth conditions for quantificational sentences being read existentially.

The semantical truth conditions suitable to the existential reading of the quantifier rely on strong semantical relations like designation, satisfaction, denotation, etc. and the statement of these conditions is treated as equivalent to the sentences for which they are the truth conditions. For instance, Frege's description of '$\exists x$' as a second level concept grows out of the attempt within his semantics of concepts and objects to give the truth conditions for the quantifiers. Russell used both the referential and the substitutional view and thus treated generalizations and their referential and substitutional truth conditions interchangeably. Danto's use of instantiation must also be commented upon. He treats 'x exists' and ' 'x' is instantiated' as equivalent, so that existence is thought of as instantiation. All these authors acknowledge that the above three sentences are equivalent. However, even if we accept them not merely as being equivalent but as providing some sort of contextual definitions, it is still a mistake to infer that 'exists' is a higher level or semantic property.

Possibly one source of this error consists in going from the equivalence:

horses exist ≡ 'horses' designates

to the identity claim:

existence = designation

Even if the equivalence were taken as providing a contextual definition, it would not sanction the identity thesis. The result of accepting the identity claim yields falsehoods involving use-mention confusions, for designation applies to expressions.

Horses (not the word 'horse') designate. (From the existential sentence plus definitional interchange by relying on the identity thesis.)

'Unicorns' (not the object, but the word) does not exist.
(From the equivalence: Unicorns don't exist ≡ 'Unicorns' doesn't designate. We detach the second component and by definitional interchange put 'exists' for 'designates'.)

It was in this same spirit of criticism that Moore pointed out that

Mr. Russell has been led to say: "Existence is essentially a property of a propositional function" .... I think this is a mistake on his [Russell's] part. Even if it is true that "Some tame tigers exist" means the same as "Some values of 'x is a tame tiger' are true" it does not follow, I think, that we can say that "exist" means the same as "is sometimes true," .... indeed, I think it is clear that we cannot say this; for certainly " 'x is a tame tiger' exists" would not mean the same as "Some tame tigers exist."[39]

A related confusion derives from the fact that we tend to ascend semantically to a linguistic level to discuss very general concepts of philosophy and especially logical constants. But it is not distinctive of either existence or quantification to talk about the terms we use for them. Such semantic ascent is of course not sufficient evidence that existence is a semantic or higher level concept.

We can now offer a comparison between Kant's view of existence and the contemporary view that existence when analyzed in terms of quantifiers is a higher level property. In a word Kant has taken the epistemological condition for knowledge of existence as equivalent to existence while the authors just mentioned take the semantical conditions for the truth of an existentially construed quantification as equivalent to existence. Parallels between epistemological or even psychological doctrines and contemporary linguistic ones have often been cited, e.g., on the one hand the principle of Humean empiricism which says that for every idea there is a corresponding impression and on the other the logical positivists' principle of verifiability. Kant spoke of existence in terms of epistemological relations between concepts and the conditions of experience. We commented upon how close this was to Berkeley's view.

x exists iff the idea of x is perceived.

x exists iff the concept of x is bound up with the material conditions of experience.

Compare these with Danto's formulation.

x exists iff 'x' is instantiated.

In Kant the relation is an epistemological one between some object's concept and its object, while in Danto the relation is a semantic one between an expression and its object.

As parallel claims, both are open to the same sorts of criticism. To start with, remember the standard objection to Berkeley, i.e., that it is not inconsistent to say that something exists and is not perceived. We have already applied this to Kant and can now do the same for Danto. There is no inconsistency in saying that something exists but that it is neither designated by nor an instance of some expression. The force of such arguments is like that of Moore's about purported definitions of goodness, namely, that in these cases we have not been provided with synonyms for the definiendum. There are also two other related criticisms. Compare 'exists' on one hand with 'to be perceived', 'to be connected with the material conditions of experience', 'to be designated', and 'to be instantiated' on the other hand. All of the latter are elliptical formulations of at least two-placed relations, e.g., _is perceived by _ , _is designated by _, etc., while 'exists' is merely a one-place property, i.e., _exists. Closely connected with this grammatical point is that in saying something exists one is not (except in special circumstances) talking about experiencing subjects and their concepts nor about linguistic entities like expressions. That is to say, the extra components of the dyadic epistemological or semantical relations are not part of what we are talking about when we say that something exists. In the linguistic case this appeared to yield the absurdity that all existential generalizations are meta-linguistic. In the epistemological case it yields the absurdity that all judgments of existence are about cognitive subjects. Recall Kant's research that what is added in an existential judgment is merely the relation to some cognitive subject.

V. 'Exists' Is Not a Descriptive Property

This claim was made by Kant and is endorsed by Danto as well as other contemporary figures. Kant said that existence does not add anything to our concept of an object. His British predecessors, Berkeley and Hume, had insisted that existence is not on a par with observable properties, i.e. one has no distinct or additional impression of an object's existence such as one might have of its being red or blue. The problem for these authors is that since objects exist, we must give an account of what the existence of an object consists of. For Kant existence in transcendental logic is a relation (or rather a modality of a relation) between a concept and an object and not a part of the concept or the object by itself. (For Danto existence is a relation between a word and an object.) The epistemological relation of a concept of a cognitive subject to an object of knowledge is the counterpart in transcendental logic to the subject-copula-predicate of general logic.

Kant assumes following Berkeley and Hume that existence is not a property of an object. If it is not a property of an object, then it would also not be a part of the concept of an object, for the object and the concept would not match. As a model (it is my impression that many who accept this claim about existence rely on some model such as the following one) for this aspect of Kant's view of existence, consider the notion of being a portrait of an existing person or of being a painting of an historical event. Take for example Stuart's painting of George Washington, or David's of some event in French history. There is no addition the artist might make to the completed paintings to indicate that it is one of an existing object. (To add something to the painting that is not in the object would defeat the purpose of being accurate, i.e., depicting the object as it is.) Nor is there some aspect of objects that the objects "wear on their sleeves" which allows one to distinguish the existing from the non-existing ones. If Stuart did a painting of Zeus or David of some event in the life of Pegasus, there might be nothing in these paintings to allow one to tell whether the objects exist or the events actually took place. A Kantian answer to the question of what makes one set of paintings different from the other would speak of the relation of the painting to the object depicted.

As striking as the above proferred analogy may be to some, there is no good reason for accepting it as a model for our analysis of existence. One of the reasons for thinking the analogy is helpful is in the view that existence is not an ordinary observable property. However, this is insufficient evidence for saying existence is not a property of objects but a relation of them to concepts or words. Logical properties (such as being self-identical or being not both [round and not round]) or theoretical properties (such as having charm) are no more or less observable than existence, yet one would hardly want to explain them by use of the above model. Part of the problem with the claim that existence is not a descriptive property is that one is not very clear as to what a non-descriptive property is. It seems incorrect to say that only observable properties are descriptive and if a property is not descriptive, if it is either meaningless, tautological, universal, modal (in the Kantian sense), higher level, or semantical, then the case for existence being non-descriptive remains to be made. On the other hand if to be 'descriptive' means to add information then to be told that George Washington existed and Pegasus did not or that black holes exist and caloric doesn't, then 'exists' is descriptive.

VI. 'Exists' Is a Disguised Quantifier

Almost all the arguments considered in this section treated existence sentences as covert quantifications. This by itself constitutes a weaker claim than any of those presupposing it. In Chapter One we traced the history of the connection between 'exists' and '∃x' and then in this chapter we remarked about another nineteenth century influence. According to the present view 'exists' is, in Ryle's phrase, a bogus predicate. Unlike a genuine predicate, sentences in which 'exists' plays the grammatical role of a predicate are translatable into sentences in which the predicate 'exists' disappears in favor of some quantifying expression. The point is sometimes made by saying that it is a grammatical but not a logical predicate. Sentences like

'Brown cows exist'

'Purple cows don't exist'

systematically mislead us into treating them like:

'Brown cows flourish'

'Purple cows don't flourish'

The former, though, are translatable in a way that reveal their "genuine" logical form:

'Some cows are brown' or 'There are brown cows' or 'There exist brown cows'

'No cows are purple' or 'It is false that there are purple cows' or 'It is false that some cows are purple' etc.

Here the quantifier '∃x', or some reading of it, performs the same role as the predicate 'exists'.

To hold that existence sentences are disguised quantifications is to claim that the existential reading is the correct one. There are difficulties for this reading which do not arise for the non-existential readings.

The first objection must by now seem to be anti-climactic. It concerns the relative expressive power of the existential reading. Sentences which are normally contingent become analytical truths and falsehoods upon equating existence and quantification. Consider the following crucial assumptions from one of Moore's arguments.

(12) '∃x', 'some', 'there are' are equated with 'there exists'—this amounts to holding that the existential reading is the correct one.

(13) Some pieces of the puzzle don't exist.

(14) There exist pieces of the puzzle which don't exist.

It is easy to imagine an occasion for using (13) to make a true statement. Think of a situation in which someone was unwittingly trying to finish a puzzle which lacked a piece. Another person might helpfully offer this information. (One could also show that (13) is not linguistically deviant by appealing to the intuitions of a native speaker.) Now if in addition we assume (12), that is to say, take the existential reading literally, then (13), an otherwise true sentence, yields the analytically false (contradictory) sentence (14). On the basis of this contradiction we could conclude (15).

(15) (12) is false, i.e., the existential reading is not correct.

Given the non-existential readings we can easily express contingent O-form existence sentences by treating 'exists' as a non-logical predicate, viz.

$(\exists x)(x$ is a piece of the puzzle and x does not exist).

A second criticism concerns a disparity between the relations of '$(\exists x)$' and '(x)' to each other and the relations between their respective existential readings. A condition for a correct definition is that the definiens and the definiendum be of the same grammatical category. This condition is met by defining one quantifier in terms the other, e.g., $(x)Fx = df \sim(\exists x) \sim Fx$. We expect that this parallelism should be preserved in all readings of the quantifier. Consider the following pairs of readings:

All	Some
For all x	For some x
Always true	Sometimes true
?	There exists an x

On the existential reading of the particular quantifier, there seems to be no clear parallel reading for the universal quantifier. Although not terribly strong, this point ought not to be dismissed and should provide motivation for re-examining the aptness of the existential reading of the quantifier.

Needless to say thinking of existence in terms of quantification was in no way part of Kant's meaning when he said that 'exists' is not a real predicate.

Summary

My goal is to separate existence from the logic of quantification. Historically this amounts to arguing for a view which appears to have been quite prominent up until the second half of the nineteenth century. Writers in the later Fregean tradition have assumed that they agreed with or even that they improved on Kant's treatment of existence. I have shown that in most cases there was no such connection. Kant's view of existence not being a real predicate was in good part an epistemological doctrine bound up with his philosophy of transcendental logic. His views on existence in general logic are derived from an older tradition linking existence and the logic of the copula. By contrast more recent account of existence have their roots in the doctrine of quantification. Even had Kant been successful in showing that 'exists' is not a real predicate it would provide no evidence for any of the views of existence in terms of quantification.

I have examined two claims:

I.   Exists is not a predicate (really a series of claims).

II.   The existential reading is correct.

and found argument for the first to be inadequate. This was so for Kant as well as for the claims about meaninglessness, analyticity, etc., which rested on the second thesis. Finally we began to consider the case where the two claims coincide, i.e., where 'exists' is not analyzed as a predicate but in terms of a quantifier. After offering some reasons for having second thoughts about the existential reading we now turn to examine it in greater detail.

Notes

1. I. Kant, *Critique of Pure Reason*, trans. N. K. Smith (London: Macmillan and Co., 1953), p. 504 (hereinafter referred to as *Critique of Pure Reason*).

2. R. Carnap, "The Elimination of Metaphysics through the Logical Analysis of Language," in *Logical Positivism*, ed. A. J. Ayer (Glencoe, Ill.: The Free Press, 1959), p. 74.

3. G. Ryle, "Systematically Misleading Expressions," *Logic and Language*, ed. A. Flew (Garden City: Anchor Books, 1965), p. 17.

4. A. J. Ayer, *Language, Truth and Logic* (New York: Dover Publications, Inc., 1946), p. 43.

5. W. V. Quine, *Mathematical Logic* (New York: Harper Torchbooks, 1951), p. 151.

6. *Critique of Pure Reason*, p. 500.

7. J. Shaffer, "Existence, Predication and the Ontological Argument," in *The Many Faced Argument*, ed. J. H. Hick and A. C. McGill (New York: The Macmillan Co., 1967), p. 228-231.

8. *Critique of Pure Reason*, p. 109.

9. *Critique of Pure Reason*, p. 185.

10. *Critique of Pure Reason*, p. 239.

11. *Critique of Pure Reason*, p. 239.

12. *Critique of Pure Reason*, pp. 239-240.

13. "I do not think there is any essential difference between existence and actuality—Kant uses the two terms as equivalent even in the present passage—but reality, although necessary to existence, is not identical with it. There are degrees of reality, but there are no degrees of existence; for a thing either exists or does not exist." This quote is from H. J. Paton, *Kant's Metaphysic of Experience*, Vol. II, (London: George Allen and Unwin, 1951), p. 357. For our purposes we can also include "reality" since Kant tells us in connection with the ontological argument that "There is already a contradiction in introducing the concept of existence—no matter under what title it may be disguised—into the concept of a thing . . . . The word 'reality' which in the concept of the thing sounds other than the word 'existence' in the concept of the predicate, is of no avail in meeting this objection." *Critique of Pure Reason*, pp. 504-505.

14. *Critique of Pure Reason*, pp. 242-243.

15. *Critique of Pure Reason*, p. 244.

16. *Critique of Pure Reason*, pp. 247-248.

17. B. Mates, "Leibniz on Possible Worlds," in *Logic, Methodology and the Philosophy of Science*, ed. B. Van Rootselaar and J. F. Staal (Amsterdam: North Holland Publishing Co., 1968). Contemporaries refer to this view as a theory of world bound individuals. See A. Plantinga, *The Nature of Necessity*, (Oxford: Oxford University Press, 1974), pp. 88-89.

18. *Critique of Pure Reason*, p. 250.

19. *Critique of Pure Reason*, pp. 251-252.

20. See M. K. Munitz, *The Mystery of Existence* (New York: Appleton-Century-Crofts, 1965), pp. 75-76.

21. *Critique of Pure Reason*, p. 502.

22. *Critique of Pure Reason*, p. 504.

23. *Critique of Pure Reason*, pp. 504-505.

24. J. Shaffer, "Existence, Predication and the Ontological Argument," in *The Many Faced Argument*, ed. J. H. Hick and A. C. McGill (New York: The Macmillan Co., 1967), pp. 228-231.

25. For an insightful view of Kant on existence see D. P. Dryer, "The Concept of Existence in Kant," *The Monist*, 50 (1966), pp. 17-33.

26. *Critique of Pure Reason*, pp. 239, 243, 250.

27. "Kant acknowledges the peculiar status of the Postulates by offering only an explanation of them, rather than the customary proof (cf. B 266). Despite its incongruity with the rest of the System of Principles, the section contains a very valuable account of the Critical interpretation of the central metaphysical terms, possibility, actuality, and necessity. As we would expect, Kant transforms them into epistemological concepts by relating them to the conditions of a possible experience." R. P. Wolff, *Kant's Theory of Mental Activity* (Cambridge: Harvard University Press, 1963), pp. 292-293.

28. *Critique of Pure Reason*, p. 505.

29. *Critique of Pure Reason*, p. 505.

30. See the translated selections from Brentano's "Psychologie vom Empirischen Standpunkt," in *Realism and the Background of Phenomenology*, ed. R. Chisholm (Glencoe: The Free Press, 1960); G. Iseminger, ed., *Logic and Philosophy* (New York: Appleton-Century-Crofts, 1968). Brentano's views on these questions were made known to English readers in an article in *Mind*: J. P. N. Land, "Brentano's Logical Innovations," *Mind*, 1 (April, 1876), p. 289.

31. M. Thompson, "On Aristotle's Square of Opposition," in *Aristotle*, ed. J. M. E. Moravcsik (New York: Anchor Books, 1967).

32. B. Mates, "Leibniz on Possible Worlds," in *Logic, Methodology and the Philosophy of Science*, ed. B. Van Rootselaar and J. F. Staal (Amsterdam: North Holland Publishing Co., 1968).

33. A. N. Whitehead and B. Russell, *Principia Mathematica to *56* (Cambridge: Cambridge University Press, 1962), p. 56.

34. For example see F. B. Ebersole, "Whether Existence is a Predicate," *The Journal of Philosophy*, 60 (August, 1963) and M. Kiteley, "Is Existence a Predicate?" *Mind*, 73 (1964), pp. 364-373.

35. For additional arguments that 'everything exists' is not true see N. Rescher's essay on the logic of existence in his *Topics in Philosophical Logic*, (New York: Humanities Press, 1968).

36. For example see B. Routley, "Some Things Do Not Exist," *Notre Dame Journal of Formal Logic*, Vol. VII No. 3 (1966). Routley introduces 'exists' and a special quantifier corresponding to it.

37. A. C. Danto, *Analytical Philosophy of Knowledge* (Cambridge: Cambridge University Press, 1968), pp. 220-221.

38. D. Welker, "Existential Statements," *The Journal of Philosophy*, 67 (June, 1970), pp. 376-388.

39. G. E. Moore, *Philosophical Papers* (New York: Collier Books, 1959), p. 122. This sort of confusion is not limited to the concept of existence. M. Levin finds the same problems arise over the concept of identity.

Philosophers who—like Frege in "Sense and Reference"—have supposed identity to be a relation between singular terms have made a similar mistake. *a* is idential to *b* when the two singular terms '*a*' and '*b*' bear the relation of codesignation to each other. So some philosophers confusedly think that *identity itself* is a relation between singular terms. [M. Levin, "The Extensionality of Causation and Causal-Exploratory Contexts," *Philosophy of Science* Vol. 43, 1976, p. 269.]

Chapter Five

Some Questions
About Explicating Existence
in Terms of Quantification

> Establishing the foundations of mathematics is not the only purpose of logic, particularly if the assumptions deemed convenient for mathematics do violence to both ordinary and philosophical usage.[1]
>
> R. Barcan-Marcus

The most serious account of the existential reading of the particular quantifier has been the one W. V. Quine has carefully worked out over a period of years. In the first part of this chapter, I will present his refinement of the thesis that the existential reading is correct, i.e., that existence claims are disguised quantifications. In the second part I will assemble criticisms of his view and try to show that he faces difficulties that the non-existential readings—particularly the substitutional one—do not.

Part I: Quine's Program for Existence and Quantification

A. The Interdependence of the Concepts of Truth and Existence

I noted above that the justification for giving a distinctive natural language reading to '∃x' depended on the semantical truth conditions for sentences containing that sign. Thus the justification for reading '∃x' existentially depends on the fact that the truth of quantified sentences involves the notion of a domain. Recall the conditions for atomic sentences and sentences with the particular quantifier.

126

val $(Pt_1, ..., t_n)$ = T iff $\langle$val $(t_1), ...,$val $(t_n)\rangle \in$ val (P)

val $((\exists x)A)$ = T iff $(\exists d)(d \in D$ & val (d/xA) = T$)$

An atomic sentence is true when the values assigned to its arguments are members of the values assigned to its predicates. The values assigned to its arguments (the arguments here are singular terms) are the individuals of the domain. The predicates are assigned sets from the domain. An "existential" quantification for an individual variable is true precisely when there is an individual in the domain which would make the appropriate substitution instance, e.g., 'Ad' true in the sense just explained for an atomic sentence. Although Quine does not present truth conditions in exactly this way, I shall nonetheless explain his position in these terms, taking care in doing so not to misrepresent it. The values of one's variables (in this case individual variables) are the members of the domain. Quine's slogan that to be is to be the value of a variable means that to be or to exist is to be a member of a domain. Quine identifies being and existence and I shall follow him in this practice while explaining his views.

It is important to note that this account makes the concepts of truth and existence interdependent. In other words, membership in a domain, i.e., existence, is the basis for computing the truth of all sentences. Thus membership in a domain,—existence—is the only basis for computing the truth of an atomic sentence, and the truth value of all other sentences is then computed from the atomic sentences.

Just as the notion of a domain is central to explaining the existential reading so is the allied notion of a semantical relation. Some of these semantical relations discussed above were designation (an expression designates or names an object), multiple denotation (an expression is true of, applies to, or denotes several objects), and satisfaction (objects satisfy an expression). In each case the relation consists of a linguistic entity on the one hand and a member or members of the domain on the other. Quine has always used such notions to justify his reading of the quantifier. In early papers such as "Designation and Existence" (1939) and "A Logistic Approach to the Ontological Problem" (1939), existential quantification is linked with naming.

Perhaps we can reach no absolute decision as to which words have designata and which have none, but at least we can say whether or not a given pattern of linguistic behavior *construes* a word W as having a designatum. This is decided by judging whether existential generalization with respect to W is accepted as a valid form of inference. A name—not in the sense of a mere noun, but in the semantic sense of an expression designating something—becomes describable as an expression with respect to which existential generalization is valid. ... instead of describing names as expressions with respect to which existential generalization is valid we might equivalently omit express mention of existential generalization and describe names simply as those constant expressions which replace variables and are replaced by variables according to the usual laws of quantification.[2]

Thus the problem of when one can existentially generalize is linked with the question of which expressions can be regarded as non-syncategorematic, i.e., as names.

To ask whether there is such an entity as roundness is thus not to question the meaningfulness of 'roundness'; it amounts rather to asking whether the word is a name or a syncategorematic expression.
    Ontological questions can be transformed, in this superficial way, into linguistic questions regarding the boundary between names and syncategorematic expressions.[3]

A recurrent theme can be discerned at this early point in Quine's writings—the explanation of variables in terms of pronouns:

Variables are pronouns, and make sense only in positions which are available to names.[4]

When Quine later developed the position that individual constants or names are a derivative type of expression that can be dispensed with, he stopped relying on the semantical relation of designating-naming. The thesis of the superfluity of names appeared first in *Mathematical Logic* (1940) and then in the better-known "On What There Is" (1948). Quine's more recent explanations of the quantifiers now appeal to notions like multiple denotation and satisfaction. In "Existence and Quantification" (1966), he has said:

Another way of saying what objects a theory requires is to say that they are objects that some of the predicates of the theory have to be true of, in order for the theory to be true. But this is the same as saying that they are the objects that have to be values of the variables in order for the theory to be true. It is the same, anyway, if the notation of the theory includes for each predicate a complementary predicate, its negation. For then, given any value of a variable, some predicate is true of it; viz., any predicate or its complement. And conversely, of course whatever a predicate is true of is a value of the variables. Predication and quantification, indeed, are intimately linked; for a predicate is simply any expression that yields a sentence, an open sentence, when adjoined to one or more quantifiable variables. When we schematize a sentence in the predicative way *"Fa,"* or *"a* is an *F,"* our recognition of an *"a"* part and an *"F"* part turns strictly on our use of variables of quantification: the *"a"* represents a part of the sentence that stands where a quantifiable variable could stand, and the *"F"* represents the rest.

Our question was: what objects does a theory require? Our answer is: those objects that have to be values of variables for the theory to be true.[5]

In his recent *Philosophy of Logic* (1970) Quine has sketched a definition of truth along Tarskian lines, relying on the relation of satisfaction.[6] A propositional function is satisfied by objects (or more precisely by sequences) found in the domain. Since Quine now holds that names are not a primitive form of expression, he would no longer accept the truth conditions presented at the start of this section. For example his semantics would not make use of arguments where *'t'* is understood as referring to a name. The relations being true of and satisfaction rely only on the primitive categories of predicates and propositional functions. However, for each of these as well as for naming, the truth of a sentence depends upon membership in a domain, i.e., existence. I emphasize once again that for Quine truth and existence are interdependent, and turn now to see how this applies to the problem of vacuous singular terms.

B. The Problem of Empty Individual Constants

In Chapter Three two requirements for any adequate theory of quantification were laid down. The first was that no particular-existential generalization be provable as a theorem. I showed at that point how Quine modified the system of *Mathematical Logic* to attain this end. The second requirement was that there be no restriction on the non-logical constants that fall under the rules of logic. The problem here was that of empty individual constants and rules such as particular-existential generalization. Recall how the presumed-to-be-true 'Pegasus is a flying horse' yielded the false (if read existentially) '($\exists x$)(x is a flying horse)'—i.e., there exists a flying horse.

Quine has presented two somewhat similar solutions to this problem. The earlier one appeared in *Mathematical Logic*, the later in "On What There Is." Common to the two is the basic idea that individual constants are dispensable and not part of our canonical notation. In both of these analyses, the troublesome sentence 'Pegasus is a flying horse' turns out to be equivalent to a false sentence, so that the inference to '($\exists x$)(x is a flying horse)' remains truth-preserving.

The solution offered in *Mathematical Logic* is a variant of Frege's chosen-object theory. Individual constants are defined in terms of definite descriptions. These descriptions are themselves analyzed in terms of class abstraction. Where the name designates a unique object it is ultimately analyzed in terms of a class expression designating a unit class. Where the name or its intermediary description is true either of more than one individual or of none the class abstract designates the null class. In other words, the object chosen for improper names and definite descriptions is the null class. 'Pegasus is a flying horse' is presumably false since 'is a flying horse' is a predicate which fails to be true of any individual in the domain (and for that matter of the null class).

An odd consequence follows from the chosen-object theory. Since the null class does exist, i.e., '($\exists x$)($x = \Lambda$)', is true, the sentence 'Pegasus exists' turns out to be true. Quine says that when we treat individual constants like 'Europe', 'God', and 'Pegasus' in terms of class abstraction "There is no question of the existence of these three entities; there is question only as to their nature."[7] I shall comment on this below.

Quine's more recent and better-known solution is an extension of Russell's theory of descriptions. According to this theory, definite descriptions are contextually defined so that they need never be taken as part of our primitive vocabulary. Quine has shown that for any individual constant an equivalent definite description can be constructed so that the former can be eliminated when we dispense with the latter.

For 'Pegasus', we already have the definite description 'the winged horse of Bellerophon'. If we did not have a description, we could easily have constructed one such as 'the unique object which pegasizes'. In canonical notation the sentence 'Pegasus is a flying horse' becomes:

$(\exists x)(x$ pegasizes & $((y)(y$ pegasizes $\supset y = x)$ & x is a flying horse)

Since the "existence" condition is false, i.e., the predicate 'is a flying horse' does not apply to anything in our domain, the entire sentence is false. Similarly, the original sentence is false; thus the argument does not constitute a counterexample to particular-existential generalization.

In summary, all simple sentences with vacuous singular subjects are treated as false in both of Quine's analyses.

C. Quantification for Grammatical Categories Other Than Singular Terms

The formula that "to be is to be the value of a variable," as well as the existential reading itself, is most plausible where the quantifiers bind individual variables. As we have seen, Quine thinks of variables as linked either to names designating the values of the variables or to the values of the variables alone. To say that for Quine the paradigm for quantifying is quantification of individual variables is certainly an understatement. He explains other authors' attempts at quantification for categories other than singular expressions by construing these in terms of quantification for singular terms. For his own purposes, he permits quantification only for singular terms i.e., individual variables. I will first outline Quine's understanding of quantification propositional positions, e.g., 'p', 'q', and predicate positions, e.g., 'F', 'G'. I then make some remarks on his own way of dealing with expressions in these positions, and its bearing on his manner of expressing the principles of logic.

One of Quine's earliest ventures into ontology was "Ontological Remarks on the Propositional Calculus" (1934).[8] If we take the existential reading literally, and assume that variables have values and the substituends for thes variables name these values, then there arises a problem in connection with propositional variables. What sort of object is the value of a propositional variable? If a sentence is the substituend for such a variable, then what sort of object does a sentence designate? Before proceeding any further I should make some presuppositions explicit. There are at least two different views which one can take about semantical relations and the place of stipulation in semantics. An extreme view, expressed by Carnap, was that it is a matter of convention which categories of expressions we consider to be designators—names.[9] According to this view sentences and predicates can be said to designate. On the other hand, there is the view which coincides with common usage, according to which only names designate. Sentences may be true or false but they do not designate. Similarly predicates may be true of or apply to objects, but they do not designate. Quine—and I—accept the latter view.

> My present objection is only against treating statements and predicates themselves as names of such or any entities, and thus identifying the 'p', 'q', etc. of truth-function theory and the 'F', 'G', etc. of quantification theory with bindable variables. . . . To withhold from general terms or predicates the status of names of classes is not to deny that there are often (or always) . . . certain classes connected with predicates otherwise than in the fashion of being named. Occasions arise for speaking of the *extension* of a general term or predicate—the class of all things of which the predicate is true. . . .
>   Similarly there is occasion in the theory of validity to speak of truth values of statements, for example in defining truth-functional validity. But there is no need to treat statements as names of these values, nor as names at all.[10]

I find it necessary to stop explicating Quine's defense of the existential reading for a moment, and make a criticism. One cannot consistently maintain that the existential reading is appropriate for all sorts of variables (to be is to be the value of a variable for any category of variables), and also hold that only singular terms are designators (individual variables have values or individual constants designate those values). Since only singular terms designate, forms of quantification other than those over singular terms cannot be explained on the existential reading. Quine himself does argue for restricting quantification to singular terms, putting aside the principle that only singular terms designate when explaining the views of others who take different styles of quantification seriously.

Returning to the propositional variables 'p', 'q', Quine notes the following alternatives. If a principle like '$p \supset (p \vee q)$' is object language, then two different objects can be taken as the values of the propositional variables. The first follows Frege's view that 'p', 'q', have one of two values—the True or the False. Thus individual sentences (the substituends for 'p', 'q') designate either the True or the False. The second choice for values of propositional variables has been propositions. Propositions here are intensional entities and would be the designata of the sentences serving as substituends for 'p' or 'q'. Another alternative which Quine mentions but rejects would consist of taking 'p' and 'q' as meta-linguistic variables having object language sentences as their values, but then '$p \supset (p \vee q)$' would no longer be an object-language principle.

Quine accords a similar treatment to quantification of predicates, e.g., $(\exists F)(Fx)$. The kind of objects that predicate variables have been said to take as values and predicate constants to designate are either the extensional entities (classes) or the intensional ones (properties-attributes). Quantification over predicate positions is a part of higher-order logic. For Quine, such quantification—existentially construed—demands values for a variable; since he is an extensionalist, he chooses sets as his values. From these two considerations, (1) higher-order logic requires values for its variables, and (2) extensionalism—it follows that for Quine, to do higher-order logic is to be involved in set theory. In his words '$(\exists F)(Fx)$' is "set theory in sheep's clothing."[11]

Quine has used these alternative treatments of divergent quantificational contexts to offer what have become influential interpretations of the history of modern logic. An example he has used repeatedly is Russell's "no class theory" of classes.[12] In *Principia*, Russell claimed to be contextually eliminating class expressions in favor of quantification over propositional functions. This was analogous to the elimination of definite descriptions in terms of quantification over individual variables. Quine interprets Russell's claim in the following way: if Russell thought that quantification over propositional functions involved only linguistic entities, open sentences or predicates, then he has confused object and meta-linguistic quantification; a propositional function as an open sentence could be the value of a meta-linguistic variable but never of an object-language one. If, on the other hand, Russell thought quantification over propositional functions involved properties-attributes, then his no-classes theory dispenses with classes in favor of more suspect intensional entities and hence fails to be a significant reduction.

So far I have only shown how Quine interprets different styles of quantification as used by other logicians. He himself, though, considers such styles of quantification mistaken and restricts quantification to the individual variables of first order logic. However, we must then ask how one can express an ontological commitment to classes or properties using only individual variables and their values. Quine's program for connecting existence and the quantifier must allow for asserting the the existence of such entities. His solution is to appeal to the traditional distinction between concrete singular and abstract terms.[13] The individual constants which can serve as substituends for the individual variables may be names of concrete individuals, e.g. 'Bossy', 'Bucephalus', or names of abstract objects like classes, e.g. 'mankind', 'red', or properties, e.g. 'triangularity', 'redness'. This is one of the most fascinating parts of Quine's program. Whatever exists is a member of the domain required for the existential reading of the quantifier. There is only one style of variable—the individual variable—and its values can be concrete objects, classes, properties, etc. This is why it is an understatement to say that Quine models quantification on quantification of individual variables. Actually all of his quantification is in respect to individual variables. The slogan "to be is to be the value of a variable" is similarly misleading. It would be more accurate to say that "to be is to be the value of an individual variable."

We have just indicated how Quine can express an existential commitment to different sorts of entities without quantifying over non-nominal positions. But such positions pose another difficulty for him, concerning the possibility of expressing the principles of logic in their full generality. Consider once again the principle of propositional logic '$p \supset (p \lor q)$'. We saw that Quine rejects the view that 'p' has values or that its substituends designate sentences. Moreover he would not like to expand his ontology solely because of propositional logic. His solution to expressing the principles of propositional logic is to regard 'p', 'q', etc. as schematic letters.[14] Schematic letters are described as being dummy expressions, or blanks in a sentence diagram. They cannot be quantified over and therein lies their essential difference from variables. The device of schematic letters also enables us to express quantificational principles like '$(x)(Fx \supset Fx)$'. The 'F' and 'G' in quantificational principles are schematic letters and not variables because we do not quantify over them. They are said to be placeholders or unbindable dummy predicates. Thus, '$x+3 = 7$' is an open sentence or propositional function, because 'x' is a free but bindable variable. '$(x)(Fx \supset p)$' is not a propositional function but a schema. Neither 'F' nor 'p' are bindable as are variables.

Schematic letters should not be confused with the meta-linguistic variables Quine uses in *Mathematical Logic*, e.g. 'ϕ', 'ψ':

A schema such as $(x)(Fx \supset p)$, –is not a name of a sentence, not a name of anything; it is *itself* a pseudo-sentence, designed expressly to manifest a form which various sentences manifest. Schemata are to sentences not as names to their objects, but as slugs to nickles. . . . The distinction which properly concerns us in the present pages, that between sentence and schema, is not a distinction between the use and mention of expressions; its significance lies elsewhere altogether. The significance of preserving a schematic status for 'p', 'q', etc. and 'F', 'G', etc. rather than treating those letters as bindable variables, is that we are thereby (a) forbidden to subject those letters to quantification, and (b) spared viewing statements and predicates as names of anything.[15]

Though schematic letters are not meta-linguistic variables, they are not part of the object language either.

Note carefully the role of the schematic letters 'p' and 'q' in the above explanations. They do not belong to the *object language*—the language that I have been explaining with their help. They serve diagrammatically to mark positions where sentences of the object language are to be imagined. Similarly, the schematic notation 'Fx' may conveniently be used diagrammatically to mark the position of a sentence when we want to direct attention to the presence therein of the variable 'x' as a free or unquantified variable. Thus we depict the form of existential quantification schematically as '$(\exists x)Fx$'. The schematic letter 'F' like 'p' and 'q' is foreign to the object language.[16]

One consequence of this use of schematic letters to express the princi-ples of logic is that none of these principles can be an object-language generalization. This follows from the fact that schemata are "foreign to the object language" and that they are not strictly speaking sentences— but only "pseudo-sentences."

D. Polemical Uses of the Existential Reading

Some of the consequences of adopting the existential reading pro-duce a body of criticism directed at higher-order logic, modal logic, doxastic logic, etc. To begin with, I have shown how the existential reading can be applied literally to non-nominal variables only if we are prepared to hold that expressions which ordinarily do not designate in fact do so. Quine takes this as indicating that disciplines like higher-order logic or quantified propositional logic should be avoided. A second, related criticism is that indulging in different styles of quanti-fication commits us to new sorts of entities. Applying Ockham's razor that entities should not be multiplied beyond necessity, we find our-selves involved in a program that limits the styles of quantification.

Other criticisms that grow out of the existential reading occur in connection with intensional contexts; recall one of Quine's criticisms of modal logic mentioned in Chapter Two.

True   (1) $\Box$ 9 is greater than 7.
True   (2)   9 = the number of the planets.
False   (3) $\Box$ The number of the planets is greater than 7.
        (4) $(\exists x) \Box (x$ is greater than 7)   from 1 by generalization

We grant that (1) and (2) are true while (3) is false. If we view (4) existentially, as Quine does, then it is read as:

(5)   There exists something necessarily greater than 7.

However, this creates a puzzle. (4) is said to follow from (1) by gener-alization. The reading for (4), i.e. (5), reveals that (4) is false, for what object is necessarily greater than 7? It cannot be the number 9, which is also the number of the planets, because that is incompatible with (3), which is also false. For Quine, the moral to be drawn is that com-bining quantification and modality reveals that modal logic is unclear and suspect.

Quine draws a similar moral about combining quantification and propositional attitudes. Consider the following case:

True   (6)   Joe believes that Mary = Mary.

        (7)   ($\exists x$)(Joe believes Mary = x) (6) Generalization

Reading (7) existentially we have:

        (8)   There exists an x such that Joe believes Mary = x.

Now imagine a situation in which there is no such person as Mary: (6) could still be true while (8) (and thus (7) too) would be false. Quine concludes from the difficulties in combining existentially-read quantification and propositional attitudes that the latter is a problematic area.

E. Quine's Meta-philosophy

In order to evaluate Quine's treatment of quantification in terms of the existential reading I must add a few words about his conception of philosophy. Quine regards his treatment of existence in terms of a special reading of the quantifier as a case of philosophical analysis or explication. Whether this explication of existence is successful depends on one's meta-philosophical opinions about the nature of such analysis.

I begin by noting one thing that philosophical analysis does *not* do, according to Quine: it does not provide us with the meaning of the expression being analyzed. Quine's skepticism about meanings leads him to eschew them in explaining philosophical analysis.

We do not claim synonymy. We do not claim to make clear and explicit what the users of the unclear expression had unconsciously in mind all along. We do not expose hidden meanings, as the words 'analysis' and 'explication' would suggest: we supply lacks. We fix on the particular functions of the unclear expression that make it worth troubling about, and then devise a substitute, clear and couched in terms to our liking, that fills those functions. Beyond those conditions of partial agreement, dictated by our interests and purposes, any traits of the explicans come under the head of "don't cares" (38). Under this head we are free to allow the explicans all manner of novel connotations never associated with the explicandum. This point is strikingly illustrated by Wiener's definition of the ordered pair.[17]

Quine chooses as his paradigm for a philosophical analysis the explication of the notion of an ordered pair. We expect our analysis to provide a substitute for those "partiuclar functions of the troublesome expression that make it worth troubling about." These conditions of partial agreement constitute the material adequacy condition for the explication. In the case of the ordered pair this condition can be stated succinctly and explicitly as

If $\langle x, y \rangle = \langle z, w \rangle$ then $x = z$ and $y = w$.

Similarly, the material adequacy condition for Tarski's explication of truth sought to capture partially certain intuitions about the correspondence theory of truth. This has become known as convention T (' 'Snow is white' is true iff snow is white' is an instance of the convention). An explication which meets the condition of material adequacy must also be formally adequate. Tarski meant by this that the formal structure of the language used in the definition has to be carefully specified. This is what Quine has in mind in his phrase "clear and couched in terms to our liking." The formal semantics provided earlier were formulated so as to meet this condition. For Quine, explication is part of philosophy, and he regards philosophy as continuous with science. The same sorts of considerations for adopting one theory rather than another are brought to bear on adopting one analysis rather than another, e.g., explanatory power, relative simplicity, etc. We must then judge whether Quine's defense of the existential reading of the particular quantifier, i.e., that quantification explicates existence, is formally, materially, and methodologically adequate. I turn now to this claim.

Part II: Is the Explication of Existence in Terms of Quantification Adequate?

The evaluation of the adequacy of explicating existence by means of quantification is complicated in at least two ways. In the first place we are not in a position to state a material adequacy condition for 'exists' as exactly as we did in the case of the ordered pair. As Quine himself acknowledges, the ordered pair is (to this extent at any rate) not a typical analysandum. Wiener's explication is not a paradigm in the sense of being perfectly typical, but rather as an ideal. It is most typical in philosophy that the material adequacy condition, "how closely we reproduce the pre-systematically available notion," is itself a bone of contention.[18] The second complication is that adequacy in the three senses of material, formal, and methodological, applies both to the explicandum 'exists' and to facets of the explicans, viz. 'some', 'there is', which are equated with 'there exists' and the technical term "$\exists x$'.

I shall offer three lines of criticism of this proposed explication, and at the same time indicate how the non-existential readings avoid these problems.

F. The Existential Reading Cannot Provide a Canon of Reasoning

Logic should leave open the question of whether sentences like 'Pegasus is a flying horse' are true. On the existential reading as construed by Quine, this issue is closed; such sentences must be false. Let us begin by indicating why logic should accommodate the truth of such sentences. Most of the people who know what Pegasus is would consider that sentence true, and to the extent that logic should provide a canon for reasoning, it should make allowance for this truth. Imagine a dispute between two students of introductory logic. A maintains that Pegasus is a flying horse and B that Pegasus is not a flying horse. After consulting a text on mythology they would conclude that A's assertion was true and B's false, and not, as on Russell's theory of descriptions, that both are false. An even more unnatural fate befalls vacuous identity sentences such as 'Pegasus is Pegasus'. Even without knowing what Pegasus is most people would consider this sentence true, yet on Russell's and Quine's view it is considered false. My point is not that these intuitions used in ordinary reasoning are ultimately sound, but merely that logic should allow for them.

So much for illustrations. I will now enunciate another adequacy condition for logic. In Chapter Three I stated two conditions for logic being free of certain assumptions. One was that there be no restrictions on the nature of the constants (*salva congruitate*) to which the principles should apply. 'Pegasus' and 'Santa Claus' are fitting substitution instances. A proposed third requirement (to be explored) is that among the atomic sentences containing vacuous terms, some can be true and some false (and, following Strawson, some neuter, i.e. neither true nor false). We want logic to be free of the assumption of the truth of any specific singular sentences. In Chapter Three I also mentioned a second requirement, viz. that there be no theorems which are particular quantifications. On the existential reading this was construed as requiring a logic free of existence assumptions. On the substitutional reading, this in effect requires that logic should not presuppose the truth of any instances of a particular generalization. The latter is of a piece with our new requirement to leave open the question of the truth or falsity of any singular sentences.

This view has been recognized by many who are concerned with the requirements for a logic's being free. Leonard, Shearn, Rescher, van Fraassen and Thomason are some of the writers who would like to allow 'Pegasus is a flying horse' to be true.[19] Strawson meanwhile has argued persuasively that some sentences with vacuous terms, e.g., 'The present king of France is bald' be considered neither true nor false. We wish to make emphatically clear that no purely formal considerations about a system of logic prohibit regarding singular sentences as true, false, or neuter. Leblanc, Thomason and van Fraassen have demonstrated that systems allowing for such sentences have the desirable properties consistency and completeness.[20]

I have argued that some singular sentences with vacuous terms should be considered true; I have also pointed out that no purely formal considerations block this requirement. It remains to show that this cannot be accomplished on the existential reading.

The essence of the semantics justifying this reading is that being a member of a domain and being an existent are the same. It was for this reason that we said that the concepts of truth and existence are interdependent here. On this view, 'Pegasus is a flying horse' must be considered false. Almost everyone agrees that Pegasus does not exist. On an existential reading as explained by Quine, it follows that 'Pegasus is a flying horse' is false since the value of this sentence is computed solely in terms of a domain of existents. For one of his approaches, such a sentence would be equivalent to another containing a definite description, i.e., 'The one and only object which pegasizes is a flying horse'. The falsity of this would be computed from that same domain of existents. Quite simply if truth and existence are interdependent then sentences about non-existents must be false.

A few comments are now in order about Frege's chosen object theory. According to Quine's version of it in *Mathematical Logic* improper descriptions always turn out to designate the null class. It is a matter of convention which object will serve as the chosen object, the only proviso being that the object be included in the domain. Three points must be made about this theory. The first is a criticism. If we take existence seriously then it simply is false that 'Pegasus exists'. If someone says that on the chosen object theory sentences like 'Pegasus is a flying horse' are really about the null class (as is 'Pegasus exists'), then we must reply that we are talking about Pegasus and not the null class, in other words, Pegasus is not identical with the null class. The real trouble with the chosen object theory is that we turn out to be talking about something other than we thought we were.

A second point about this theory is that it actually reinforces the view that from the standpoint of formal logic it does not matter which singular sentences are true. On Quine's convention, 'Pegasus is a class' turns out to be true. If Mount Everest had been our chosen object, then 'Pegasus is the highest mountain' would be a true sentence. Theoretically it would seem possible to choose the objects in just such a way as to make singular sentences have just the truth values we want. In other words, the chosen-object theory furnishes a clue to the fact that a formally correct system of logic need make no commitment as to the truth value of singular sentences.

In Chapter Three I noted Leblanc's use of inner and outer domains as part of a semantics for the existential reading that would sanction the truth of 'Pegasus is a flying horse'. The truth conditions for an atomic sentence indicated that it would be true if the argument designated a member of either the inner or the outer domain. Pegasus would be allowed membership in the outer domain. An "existential" generalization would be true, however, only because of the inner domain.

$$\text{val } (Pt_1,...,t_n) = \text{T iff } \langle \text{val } (t_1),...,\text{val } (t_n) \rangle \in \text{val } (P)$$
>   where P and $t_1,...,t_n$ have their valuation with respect to both the inner and the outer domain.

$$\text{val } (\exists x)A) = \text{T iff } (\exists d)(d \in D_i \ \& \ \text{val } (d/xA) = \text{T}).$$

This is like the chosen object theory except that here we have a separate domain of chosen objects. The trouble with this approach is that the rationale for the existential reading consists of linking the notions of quantification and membership in a domain. If one accepts (as Leblanc did here) this rationale for the inner domain, then it is arbitrary not to do so for the outer domain as well. A further consequence of this view is that the now true Pegasus sentence is no longer equivalent to one that it should be equivalent to on the theory of descriptions, viz. 'The one and only object that pegasizes is a flying horse'. The latter would be false because the "existence" condition would not be met since it is computed solely from the inner domain. Hintikka does not seem to have noted that this same problem applies to his work. He regards Moore's 'This is a hand' as true and at the same time denies that it implies an existential sentence. Quine is consistent on this point—both the Pegasus sentence and the one with the counterpart description are false. It can also be consistently held that the Pegasus sentence is true and that individual constants can be eliminated in terms of equivalent counterpart descriptions if one adopts the substitutional reading. In this last case, the above sentences are still equivalent because they are both true. On the substitutional reading, there is no "existence" condition to be met in order for the sentence with the definite description to be true.

A possible moral to be drawn is that the truth or falsity of all singular sentences, especially those with vacuous subjects, should not be computed from considerations about a domain. This is precisely what we did when I stated the non-referential substitutional truth conditions for atomic sentences. We said that an atomic sentence has a truth value without any reference to whether or not it designated a member of a domain. Recall that

$$\text{val } (A) = \text{T iff } A \in S_t;$$

in other words, A is true if and only if A is a member of the true state description. Here the truth of an atomic sentence is entirely divorced from considerations of existence. Truth may be broader than existence.

Until now I have shown that domains as domains of existents will not allow for the truth of certain sentences. The question must be raised as to why a domain should be limited to existents and not expanded to the broader category of beings? If we wish to consider 'Pegasus is a flying horse' as true, while Pegasus does not exist, then why not consider Peagsus as a non-existent being? Let beings, and not merely existents, be the objects that make up a domain. We might take Quine's formula *au pied de la lettre* ("to *be* is to be the value of a variable") and not—as he does—as 'to *exist* is to be the value of a variable'. Here, too, we note that there are no purely formal considerations which mitigate against our following this policy. The truth condition for the existential reading is equally open to being used in this way:

$$val \left((\exists x)A \right) = T \text{ iff } (\exists d)(d \in D \ \& \ val \ (d/xA) = T)$$

Quine believes that we cannot distinguish being and existence. However we can succeed in indicating that there is a difference between them. I noted this point above with regard to the following examples.

(9)   Pegasus is a flying horse.
(10) There *is* a flying horse.
(11) There *exists* a flying horse.

By contrasting (10) and (11) we can see that there is a difference. Granted that (9) is true (10) remains true while (11) is false. Though we wish to distinguish being and existence, we nonetheless find that there is something *ad hoc* about the idea of a domain of beings.

To believe that domains are populated by beings as well as existents amounts to abandoning the existential reading for what we called the neutral 'there is' reading. I will however repeat a reason for not adopting this view of domains as populated by beings. As I have indicated there appears to be no purely formal reason for rejecting the view, but there is a methodological objection against it. The neutral 'there is' reading can be justified by a mixed semantics. In essence it consists of using the substitutional condition to account for truths about non-existent beings, such as the one about Pegasus, and using membership in a domain to account for truths about existent beings.

$$val \ (Pt_1,...,t_n) = T \text{ iff } \langle val \ (t_1),..., val \ (t_n) \rangle \in val \ (P) \text{ or}$$
$$A = Pt_1,...,t_n \ \& \ A \in S_t.$$
$$val \left((\exists x)A \right) = T \text{ iff } (\exists d)(d \in D \ \& \ val \ (d/xA) = T) \text{ or}$$
$$(\exists s)(s \in V \ \& \ val \ (s/xA) = T)$$

Our objection to positing beings is that this seems to be an *ad hoc* device to account for the truth of Pegasus-type sentences in terms of a domain. If there are no additional reasons for positing these beings as members of a domain, then the methodologically correct and simpler procedure would be to treat the truth of such sentences as *sui generis*. This of course is what we did in the substitutional reading and in the mixed semantics for the neutral 'there is' reading.

G. The Existential Reading Cannot Be Used to Express the Kinds of Generality We Expect of the Quantifiers

This section centers on the inability of the existential reading to do justice to quantification for grammatical categories other than that of singular terms. I will show that according to three different intuitions about the role of quantification, (viz., expansions, pronouns, and formal principles), the notion of quantification of, say, a predicate, is as natural as that of a singular term. The difficulty of giving a plausible account of these divergent styles of quantification constitutes a case against referential quantification. At the same time the plausibility of such quantification on the non-referential substitutional view is a telling argument for its being more basic. Some have tried to argue that non-substitutional referential quantification is basic and that substitutional quantification is, at best, derivative. The force of my argument is just the reverse: since substitutional quantification is suited naturally to quantification for diverse grammatical categories, while referential quantification is natural only for the category of singular terms, the evidence is that the latter view of quantification is a special case of the former. It is somewhat ironic that those who have argued against the adequacy of substitutional quantification have dwelled on the cases where there may be unnameable objects, while ignoring the complete inappropriateness of non-substitutional referential quantification for non-naming positions.

One of the clearest intuitions as to what is expected of '(x)' and '$(\exists x)$' is that they should be analogous to conjunction and alternation. Indeed, in the finite case the quantifiers are probably definable in terms of these logical connectives. If it is natural to think of '$(x)Fx$' and '$(\exists x)Fx$' along the lines of

'Fa & Fb & etc.' and 'Fa v Fb v etc.'

then surely by extension the same applies to thinking of '$(\phi)\phi a$' and '$(\exists \phi)\phi a$' as

'*Fa & Ga* & etc.' and '*Fa* v *Ga* v etc.',

and to '$(p)(p \supset p)$' as `

'John is tall $\supset$ John is tall & Henry is tall $\supset$ Henry is tall & etc.'

and to '$(\exists f)(p\ f\ p)$' as

'$(p \supset p)$ v $(p\ \&\ p)$ v etc.'.

A second intuition about quantifiers and variables is one brought to prominence by Quine himself. His intuition is that the role of variables is analogous to that of pronouns in ordinary language; pronouns are similar to variables in that they are the vehicles of cross reference in natural languages. Quine has constructed examples showing how the cross reference achieved in formulas with individual variables can also be achieved (though it is much more cumbersome) with pronouns such as 'it', 'the former', 'the latter', 'the first', and 'the second'.[21] The counterpart of the formula

$$(x)((y)(y<x \text{ v } y>x))$$

(assuming that the variables have only numbers as their values), could be expressed in ordinary language using 'the former' and 'the latter' as

Whatever number you may select, it will turn out, whatever number you may next select, that the latter is less than, equal to, or greater than the former.

This might also be expressed by using 'it' with subscripts.

Whatever$_1$ first number is chosen and whatever$_2$ second number is chosen, it$_2$, the second, is greater than, equal to or less than it$_1$, the first.

If this analogy is helpful, then we are very naturally led to the idea of variables for yet other grammatical categories, since pronouns serve equally well as devices of cross-reference to expressions like sentences and predicates.[22] Consider the following examples. 'If the government recalls the ambassador it means trouble'. Here the 'it' in the consequent cross-refers back to the sentence 'The government recalls the ambassador'. One could thus conclude that there is something such that if it happens it means trouble. This might be symbolized as

'$(\exists p)(p \supset p$ means trouble$)$'

For predicates consider the following:

'Now while Newton was a bachelor and could concentrate in his work, the former is true of John while the latter is not'.

One might infer from this that something is true of Newton but not of John. That is to say

'$(\exists \phi)(\phi$ Newton & $\sim\phi$ John$)$'

Our last reason for taking quantification for different types of variables seriously is that principles of logic can be enunciated for them on a par with those applying to quantification and individual variables. Rules of substitution and other quantification principles can be formulated for any grammatical categories. For simplicity's sake consider the following principle for distributing the universal quantifier over a conditional:

$(x)(Fx \supset Gx) \supset ((x)Fx \supset (x)Gx)$

$(p)(p \supset p) \supset ((p)p \supset (p)p)$

The above considerations provide the following adequacy condition for any account of quantification: quantification should not be restricted to any one grammatical category. Earlier I showed how, on the existential reading, quantification over sentential positions necessitated treating these variables as having values. In other words it required treating the substituends of the variables, in this case sentences, as designating, i.e. naming the objects serving as values. So sentences were said to naime either propositions or one of two objects, the True or the False. Further categories such as sentential connectives '$\supset$', '&', '$\vee$', etc., would, because of quantifications like '$(\exists f)(p \, f \, p)$', also have to be construed as names. No one has ever commented as to what they would name. It is at the very least unnatural to consider 'John runs', 'runs' and 'and' as names, and more likely simply false. Furthermore, there do not appear to be any theoretical advantages that would accrue if we adopted this as a strange but possibly useful convention. Indeed, since Ockham's razor provides us with an argument for *not* thinking of logic as requiring special entities which "non-names" denote, we can dispense with them as on the substitutional view.

Yet another consideration along these lines is the question of whether one can express principles of logic as general truths on the existential reading. Church has given some reasons why we should want to state logical principles in this way.

That logic does not therefore consist merely in the methatheory of some object language arises in the following way. It is found that ordinary theories, and perhaps any satisfactory theory, of deductive reasoning in the form of a methatheory will lead to analytic sentences in the object language, i.e., to sentences which, on the theory in question, are consequences of any arbitrary set of hypotheses, or it may be of any arbitrary non-empty set of hypotheses. These analytic sentences lead in turn to certain generalizations; e.g., the infinitely many analytic sentences A v $\sim A$, where A ranges over all sentences of the object language, lead to the generalization p v $\sim p$, or more explicitly $(p)(p$ v $\sim p)$; and in similar fashion $(F)(y).(x)$ F $(x) \supset F(y)$ may arise by generalization from infinitely many analytic sentences of the appropriate form. These generalizations are common to many object languages on the basis of what is seen to be in some sense the same theory of deductive reasoning for the different languages. Hence they are considered to belong to logic, as not only is natural but has long been the standard terminology.

Against the suggestion which is sometimes made from a nominalistic motivation, to avoid or omit these generalizations, it must be said that to have, e.g., all the special cases A v $\sim$A and yet not allow the general law $(p)(p$ v $\sim p)$ seems to be contrary to the spirit of generality in mathematics, which I would extend to logic as the most fundamental branch of mathematics. Indeed such a situation would be much as if one had in arithmetic $2+3 = 3+2$, $4+5 = 5+4$, and all other particular cases of the commutative law of addition, yet refused to accept or formulate a general law, $(x)(y).x+y = y+x.$[23]

As an illustration of a system of logic in which principles are expressed as general truths, consider an axiom and thesis of Lesniewski's protothetic.

Axiom: $(f)(q)[f(p)(p \supset p) \supset (f(p)(p) \supset fq)]$

This thesis corresponds to one of the paradoxes of material implication:

$(p)[(p \supset (p)p) \supset \sim p]$

For Quine, principles such as $p \supset p$ or $(x)(Fx \supset Fx)$ are not sentences, but schemata; 'p' and 'F' are schematic letters. First of all recall that schemata are not part of the object language. In a passage quoted earlier, Quine informs us that they are "foreign to the object language." Secondly, schemata are not strictly speaking sentences, but only "pseudo-sentences." From these considerations it follows that there can be no general object language truths of logic. Indeed, solely from the fact that schemata are not sentences it follows that there are no such truths of logic since, strictly speaking, only sentences are true or false.

A number of other questions can also be raised here. Schematic letters are distinguished from both meta-linguistic as well as object-language variables. Furthermore, they are neither names of sentences nor sentences.

A schema such as '$(x)(Fx \supset p)$', . . . is not a name of a sentence, not a name of anything; it is itself a *pseudo-sentence*. . . . Schemata are to sentences not as names to their objects, but as slugs to nickels.[24]

Even if a clear syntax and semantics for the notion of a schema were provided (Quine does not appear to have done so), would it not still be simpler to get on with variables, object-language or otherwise, rather than introduce a new type of expression? Quine and those who see a genuine reduction in dispensing with individual constants ought to be impressed by a method for dispensing with schemata.

If schemata are not part of our languages—object language or meta-language—then how do we understand them? By what means do we decide that the string '$p \supset p$' is a schema but that the strings '$p \supset \supset$' or '$\supset q$' are not? How do we explicate the apparently semantic property of being valid which some schemata have? Is the answer to such questions that we need to mimic the ordinary syntactical and semantical rules of language in explaining schemata, and still deny that schemata are part of any language? To do so seems like making a distinction where there is no difference—the strings obey the ordinary rules for being in a language but are not.

Another alternative is to say that schemata are in a language and have their own rules for well-formedness and for having semantical properties. However, to follow this course is to forego the wisdom of Ockham's razor. It would be simpler to make do with substitutional variables and the existing syntactic and semantic accounts of them instead of introducing schematic letters. Indeed, if one introduces schemata into one's language they appear to function in the same way as substitutional variables.

Yet another consideration stems from Quine's own conception of logical truths as being continuous with those of science. This continuity thesis clashes with the semantic ascent that takes place when one goes from the object language truths of mathematics and physics to the schemata (no longer truths) of logic.

Summarizing, I note that on the existential reading, as utilized by Quine, we cannot quantify with respect to different grammatical categories, but on the substitutional reading we can. Furthermore, on the existential reading there appears to be no way in which we can have general truths of logic in the object language.

H. The Existential Reading Is Not Philosophically Neutral and Creates Special Problems of Its Own

My last criticism concerns the correctness of equating 'some', 'there is', '∃x' and 'there exists'. I cannot claim that a contradiction results from adopting the existential reading, but I can show that it is not topic neutral and that it creates special problems which do not exist on the substitutional reading. I shall mention problematic areas such as modal and doxastic logic, and higher-order logic, as well as the unproblematic case of propositional logic. I have already mentioned that problems occur when we combine "existential quantifiers" with intentional concepts. Some of these problems do not arise if we adopt ‘a non-existential reading. Certainly there still remain problems concerning modal and doxastic logic.

Next consider the case of higher-order logic, i.e., quantification of predicate positions. For an extensionalist, higher-order logic becomes associated on the existential reading with an ontology of sets or classes. Again I acknowledge that there are serious problems about higher-order logic, e.g., incompleteness in an important sense and a Russell-type paradox for predicates. But these are essentially logical and not necessarily ontological problems.

In each case, an ontological prejudice is created about the problematic area. Even in the otherwise non-problematic case of the propositional calculus, special ontological problems come into being on the existential reading, viz. an undesirable ontology of either the intensional entities, propositions, or the extensional but mysterious objects, the True and the False.

On the existential reading, an ontological stigma becomes attached to these areas of logic. Kripke, in the course of delivering a sermon on bad habits philosopher-logicians fall prey to, provides a moral that applies here: *"Philosophers should not confuse their own particular philosophical doctrines with the basic results and procedures of mathematical logic."*[25]

By linking the logical with the ontological-metaphysical, our metaphysical prejudices provide criteria for making otherwise purely logical decisions. For instance, the ontological prejudice against the existence of sets or properties in addition to that of individuals becomes a reason for avoiding higher order logic; certainly the writing of the history of recent philosophy is colored by this approach. Quine and his followers until quite recently never considered any alternatives to the existential reading. Whenever they encounter quantification, they ask what the values of the variables are. Consider what happens in the cases of defining identity, Ramsey's treatment of theoretical predicates, and Russell's treatment of classes. No one from this school puts much stock in defining identity in higher order logic, e.g., $x = y = \mathrm{def}$ $(\phi)\,(\phi x \equiv \phi y)$, because of the ontic import of the variable 'ϕ'. Similarly, Ramsay's suggestion for quantifying over what might be called theoretical predicates poses special ontological problems on the existential reading. This is quite ironic since Ramsay himself appears to have regarded quantification of predicates in a substitutional way.

As a final example, consider Russell's account of class expressions as incomplete symbols which can be defined away contextually. This is Russell's no-classes theory. As early as November, 1905, in "On Some Difficulties in the Theory of Transfinite Numbers and Order Types" Russell mentioned three ways of treating classes so as to avoid various problems. One of these was the no-classes theory; in a note appended to that paper on February 5, 1906 Russell said:

From further investigation I now hardly feel any doubt that

the no-classes theory affords the complete solution of all the difficulties stated in the first section of this paper.[26]

Later (April 24, 1906), the London Mathematical Society heard Russell give a paper "On the Substitutional Theory of Classes and Relations."[27] Here Russell develops more fully the idea of classes as incomplete symbols, but, because of some difficulties in it, the paper was never published. The no-classes theory became the view propounded in *Principia Mathematica* and then repeated in the *Introduction to Mathematical Philosophy*.

Russell's treatment of classes (and the attendant paradoxes) developed out of his treatment of definite descriptions. In his paper "On Denoting" (probably thought out by the late spring of 1905), 'the present king of France' is treated as having no meaning in isolation, i.e., as an incomplete symbol, and as yielding paradoxes if taken as a complete symbol. In "On the Substitutional Theory of Classes and Relations," class expressions are compared to definite descriptions: both can be defined away contextually. However classes are said to be even more shadowy entities than the present king of France:

"the number one" is even more shadowy than "the present king of England"; for every possible statement about a real entity remains significant if made about the present king of England, whereas only certain kinds of statements can be made about the number one. "The number one is bald" or "the number one is fond of cream cheese" are, I maintain, not merely silly remarks, but totally devoid of meaning. In fact, all the statements about the number one that strike one as nonsensical are nonsensical in the strictest sense of the word, that is, they are phrases which do not express propositions at all. The same thing holds, on the theory in question, concerning all classes and relations.[28]

Russell seems to be saying here that class expressions and definite descriptions are both incomplete symbols, but that while improper descriptions yield only false sentences class expressions can in addition result in type errors. In order to make the comparison between these two more apparent I shall indicate (somewhat inaccurately, e.g., omitting type considerations) their contextual definitions.

For definite descriptions:

$f\ x\psi x = \text{def.}\ \underbrace{(\exists y)}\ \underbrace{[(x)(\psi x \supset x = y)]}\ \&\ \underbrace{[fy]}$

$\quad\quad\quad\quad$ "existence" uniqueness predicative
$\quad\quad\quad\quad$ condition clause clause

For classes:

$f\ \hat{z}(\psi z) = \text{def.}\ \underbrace{(\exists\phi)}\ \underbrace{[(\phi x \equiv \psi x)]}\ \&\ \underbrace{f\{\phi\hat{z}\}]}$　　　See *Principia* *20.01

$\quad\quad\quad\quad$ "existence" uniqueness predicative
$\quad\quad\quad\quad$ condition clause clause

Now bearing in mind the higher-order quantification, i.e., '$(\exists\phi)$' in the definition of classes, consider Quine's comments.

> Russell ([2], [3], *Principia*) had a no-class theory. Notations purporting to refer to classes were so defined, in context, that all such references would disappear on expansion. This result was hailed by some, notably Hans Hahn, as freeing mathematics from platonism, as reconciling mathematics with an exclusively concrete ontology. But this interpretation is wrong. Russell's method eliminates classes, but only by appeal to another realm of equally abstract or universal entities—so-called propositional functions. The phrase 'propositional function' is used ambiguously in *Principia Mathematica*; sometimes it means an open sentence and sometimes it means an attribute. Russell's no-class theory uses propositional functions in this second sense as values of bound variables; so nothing can be claimed for the theory beyond a reduction of certain universals to others, classes to attributes. Such reduction comes to seem pretty idle when we reflect that the underlying theory of attributes itself might better have been interpreted as a theory of classes all along, in conformity with the policy of identifying indiscernibles.[29]

Quine never considered the possibility of construing substitutionally the higher-order quantification in the no-classes theory. While it is true that Russell frequently made use-mention errors, and that he wished to accommodate intensional entities, it is nonetheless also true that he propounded a substitutional view of quantification. Russell never seemed to notice that he held two different readings of the quantifiers and that they could diverge. We may conclude that on the substitutional reading the no-classes theory could furnish a significant reduction.

Notes

1. R. B. Marcus, "Extensionality," in *Reference and Modality*, ed. L. Linsky (Oxford: Oxford University Press, 1971).

2. W. V. Quine, "Designation and Existence," in *Readings in Philosophical Analysis*, ed. H. Feigl and W. Sellars (New York: Appleton-Century-Crofts, 1949), pp. 49-50.

3. W. V. Quine, "A Logistic Approach to the Ontological Problem," *Ways of Paradox* (New York: Random House, 1966), p. 64 (hereinafter cited as *Ways of Paradox*).

4. Quine, *Ways of Paradox*, p. 65.

5. W. V. Quine, *Ontological Relativity and Other Essays* (New York: Columbia University Press, 1969), pp. 95-96.

6. W. V. Quine, Philosophy of Logic (Englewood Cliffs, N. J.: Prentice Hall, 1970), chapter 5 (hereinafter cited as *Philosophy of Logic*).

7. Quine, *Mathematical Logic*, p. 150.

8. Quine, *Ways of Paradox*.

9. See Carnap, *Meaning and Necessity* (Chicago: University of Chicago Press, 1956), p. 7.

10. W. V. Quine, "Logic and the Reification of Universals," *From a Logical Point of View* (New York: Harper Torchbooks, 1961), pp. 112-115 (hereinafter cited as *From a Logical Point of View*).

11. Quine, *Philosophy of Logic*, pp. 64-65.

12. See Quine, "Whitehead and the Rise of Modern Logic," *Selected Logical Papers* (New York: Random House, 1966), pp. 21-23; "Russell's Ontological Development," *Journal of Philosophy*, 63 (November, 1966), pp. 659-661; J. van Heijenoort, ed., *From Frege to Gödel: A Source Book in Mathematical Logic 1879-1931* (Cambridge: Harvard University Press, 1963), pp. 249-254.

13. Quine, *From a Logical Point of View*, pp. 112-117; also Quine, *Methods of Logic* (New York: Holt, Rinehart and Winston, 1963), pp. 203-204.

14. Quine, *From a Logical Point of View*, pp. 107-117; see also Quine, *Methods of Logic*, index entry under "Schema."

15. Quine, *From a Logical Point of View*, p. 111.

16. Quine, *Philosophy of Logic*, pp. 24-25.

17. W. V. Quine, *Word and Object* (New York: Wiley and Sons,

1960), pp. 258-259.

18. A. Church, "Ontological Commitment," *Journal of Philosophy*, 55 (1958), p. 1012.

19. H. S. Leonard, "Essences, Attributes and Predicates," Presidential Address for the 62nd annual meeting of the Western Division of the American Philosophical Association at Milwaukee, Wisconsin, April-May, 1964, pp. 29-30, 51; M. Shearn, "Russell's Analysis of Existence," *Analysis*, 11 (1950), p. 127; N. Rescher, *Topics in Philosophical Logic* (New York: Humanities Press, 1968), pp. 152, 159; B. van Fraassen, "Presuppositions, Supervaluations, and Free Logic," and R. Thomason "Modal Logic and Metaphysics," in *The Logical Way of Doing Things*, ed. K. Lambert (New Haven: Yale University Press, 1969), pp. 89-90, 129.

20. See H. Leblanc and R. H. Thomason, "Completeness Theorems for Some Presupposition-free Logics," *Fundamenta Mathematicae*, 62 (1968), pp. 126-164; also B. van Fraassen, "The Completeness of Free Logic," *Zeitschrift für Mathematische Logik und Grundlagen der Mathematik*, 12 (1966), pp. 219-239; and B. van Fraassen, "Singular Terms, Truth Value Gaps and Free Logic," *Journal of Philosophy*, 63 (1966), pp. 481-495.

21. Quine, *Mathematical Logic*, pp. 65-71.

22. This line of criticism as well as most of the examples are from H. Hiz, "Referentials," *Semiotica*, 12 (1969), pp. 136-166. Hiz however (pp. 147-148) considers the analogy between variables and pronouns to be somewhat misleading. Here are some of his reasons. (1) The comparison of variables and pronouns does not take into account other referentials such as classifiers, e.g., 'Jean and Peter went to the movies. The man paid for the tickets'. (2) Quine treats the nominal category as the only vehicle of generality and moreover confines himself to pronouns that cross-refer to singular terms. (3) Many referentials are more like constants than variables.   For example, in 'John took his book', 'John' is a constant and so is 'his'.

23. A. Church, "Mathematics and Logic," in *Logic, Methodology and Philosophy of Science*, ed. E. Nagel, P. Suppes, and A. Tarski (Stanford: Stanford University Press, 1962), pp. 181-182. For views to the contrary see R. M. Martin, *Belief, Existence and Meaning* (New York: New York University Press, 1969), pp. 18-21.

24. Quine, "Logic and the Reification of Universals," *From a Logical Point of View*, p. 111.

25. S. Kripke, "Is There a Problem about Substitutional Quantification?" in G. Evans and J. McDowell, ed. *Truth and Meaning* (Oxford: Clarendon Press, 1976), p. 408.

26. "On Some Difficulties in the Theory of Transfinite Numbers and Order Types," *Proceedings of the London Mathematical Society*, 4 (1906), pp. 29-53.

27. This paper is mentioned in the back of the *Proceedings of the London Mathematical Society*, 4 (1906), p. viii. The paper was communicated to the society but never published. A Xerox copy of Russell's original manuscript plus letters relating to it are available from the Russell Archives at McMaster University.

28. "On the Substitutional Theory of Classes and Relations," Xerox copy, p. 2.

29. Quine, "Logic and the Reification of Universals," *From a Logical Point of View*, pp. 122-123.

Chapter 6

Conclusion

The aim of this monograph has been first, to cast doubt on the existential reading of the quantifier and attendant misconceptions about existence, and second, to argue for alternative non-existential readings (in particular for the primacy of the substitutional view). My method has been both historical and analytical; it is no secret that the existential reading has almost attained the position of an orthodoxy in philosophical circles. In fact it is virtually the "established" way of treating both existence and the particular quantifier. By presenting a history of the acceptance of this view, I hope to have undermined it. In Chapter One, I traced the linkage of existence and quantification from Frege. Later I attempted to get a larger perspective of that period in two ways. I indicated that the substitutional reading was part of the history of recent logic even though it has been ignored for the most part. Indeed, the last section suggests that the history of recent philosophy should be re-examined with non-existential readings in mind. A second way in which I sought to undermine the existential view was by showing that historically it constitutes an anomaly, as there is no clear precedent prior to Brentano and Frege for linking quantification and existence.

In Chapter Four I analyzed misconceptions attendant upon the view that existence is a matter of quantification, offering refutations of all but the claim for the correctness of the existential reading itself. In Chapter Five I backed up this claim and presented three criticisms directed at Quine's explication of existence in terms of the particular quantifier. The first two center on the issue of material adequacy, i.e. the success of the explication with regard to our intuitions about the notions involved. The first criticism was that the existential reading does not provide a canon for our reasoning about non-existent objects. The second concerned the adequacy of the existential reading in accounting for different styles of quantification. The third criticism attempted to put into perspective the fact that the existential reading is not philosophically neutral and creates problems which would not exist otherwise. For each of these criticisms I have indicated that the problems don't arise on non-existential readings. Moreover among these non-existential readings, the substitutional takes a position of primacy. We have seen that the semantics for the substitutional reading seems to be the only one to do justice to the truth values of vacuous singular sentences. Furthermore, quantification of diverse grammatical categories is a natural consequent of the substitutional view. Considerations such as these lead me to the conclusion that there should be more investigation and greater employment of the substitutional view. Logic could then be free of ontology; but what then for ontology? It is time to turn and discuss questions of existence in metaphysics proper, aided of course by a philosophically neutral logic. To begin with, if existence is not a matter of quantification, then what is it?

Appendix:
The Copula and Existence

In this appendix I will focus on Lesniewski's conception of existence and his system of logic. It is divided into three sections:

A. Two traditions of existence as a logical concept.
B. Is quantification in Lesniewski's systems to be construed substitutionally?
C. In what sense is Lesniewski's logic a free logic?

In the first of these I will indicate how Lesniewski's views on existence and quantification differ from those such as Quine's which grow out of the *Principia* tradition. In the second section I will offer an interpretation of Lesniewskian quantification (*contra* Quine, Küng, and Canty) as being both substitutional and referential.

A. Two Traditions of Existence as a Logical Concept

The claim that 'exists' is not a real predicate can be construed in a number of ways. A minimal claim is that, unlike that of real predicates, the function of 'exists' can be accomplished by using purely logical concepts. In the Frege-Russell-Quine tradition, one appeals to the particular, or so-called "existential" quantifier. In the Lesniewskian tradition (as in Kant, Leibniz, and probably other earlier figures), one appeals to a form of the copula 'est', 'jest', i.e., Lesniewski's epsilon 'ϵ'. (To avoid confusion with other uses of 'ϵ', I will, where necessary, follow Kotarbinski in using the Latin 'est' as a rendering of Lesniewski's

159

epsilon.) Both traditions treat sentences of the type '——exists' as systematically misleading and unlike those such as '——is brown'. The latter type of sentence contains 'is brown,' a real predicate. Existence sentences in unregimented natural languages can be regimented so as to yield sentences in which the function accomplished by the 'exists' of English is now performed by a logical constant. More succinctly put, existence sentences can be contextually defined in terms of other sentences in which 'exists' no longer appears. This is not the case for sentences with real predicates such as 'cows are brown'. For the quantificational tradition, 'cows exist' is regimented as '(∃x) (x is a cow)'. For Lesniewski, 'cows exist' is regimented as '(∃a) (a est cow)'.

However, at least two questions arise. Are the two traditions on a par as to this minimal claim that 'exists' is to be explicated in terms of purely logical concepts and where, if at all, do the two traditions differ? With the first of these questions, a problem arises as to the grounds for regarding Lesniewski's epsilon as a logical concept. Both traditions assume that the quantifiers are logical constants. However, one might want a justification for similarly classifying this special form of the copula.

Quine in his *Philosophy of Logic* while discussing the question of whether '=', the identity sign and 'ε', the epsilon of set theory, are logical constants (the latter must not be confused with Lesniewski's 'ε', for among other things it is not transitive while the 'ε' of Lesniewski's system of ontology is), presents the following as desiderata for an expression's being a logical constant:

1) Topic Neutrality. The predicate should not presuppose a special or restricted ontology. Thus '=' applies to any objects, e.g., physical objects or sets, while 'ε' of set theory does not. 'ε' requires sets exclusively for one of its relata.

2) The consistency and completeness of their theories. Identity theory differs in this respect from set theory as well, in that it is consistent and complete, whereas set theory is not.

3) The plausible definability or reduction of the constant in question in terms of the language of quantifiers and connectives. This criterion is also satisfied by identity, but not by the membership sign of set theory.

Hence, according to the above considerations identity qualifies as a logical constant while membership in a set does not. Lesniewski's 'ε'—'est' is assuredly topic neutral, as nouns for any category of objects (of any ontological category, that is) can be combined with an appropriate 'est' to yield a true statement. The system of logic, "ontology," in which 'est' occurs as a primitive has been shown to be consistent relative to an underlying system of logic (protothetic). Protothetic is an extended propositional calculus enriched by quantifiers for diverse grammatical categories. There is as yet no completeness proof for ontology. However, to the extent that one places certain constraints on the language of ontology (in effect, makes assumptions that are common to *Principia* type languages), it is likely that ontology can be proved to be complete.[1] Last of all, so far as I know, no one has ever claimed that 'est' is reducible to the concepts of quantification and truth functional operations.

In addition to the minimal claim that existence sentences be thought of as particular generalizations, the quantificational tradition has been associated with other more ambitious claims—that 'exists' is an anomalous, if not meaningless, predicate (this assertion is usually reserved for singular existence claims), or that 'exists' is an analytic, a universal, or a higher order predicate. Such claims are not part of the Lesniewskian tradition.

B. Is Quantification in Lesniewski's Systems to Be Construed Substitutionally?

Quine, in some recent work, has stated that Lesniewskian quantification is substitutional and not referential. Quine's reason is that Lesniewski has quantifiers for diverse grammatical categories (e.g. for sentences and connectives in protothetic, and, of special interest to us here, for nouns in elementary ontology), and yet a philosopher-logician of a nominalist cast of mind, such as Quine presumes Lesniewski to be, would not wish to be committed to the existence of abstract objects such as the sets nouns might name. Another reason for this conjecture, although only implied by Quine, is that for Lesniewski the particular quantifier does not have the force of existence. For example, reflect on how the sentence 'Something doesn't exist' fares in the Lesniewskian and Quinean traditions. For Lesniewski it is rendered as '$(\exists a)\ (\sim(\mathrm{ex}(a)))$', i.e., '$(\exists a)\ (\sim(a$ exists$))$', and is a contingent truth: some object a doesn't exist, given the truth of a substitution instance such as 'Pegasus doesn't exist', i.e., '$\sim(\mathrm{ex}(\mathrm{Pegasus}))$'. Quine, on the other hand, has claimed in *Mathematical Logic* that "To say that *something* does not *exist*, or that there *is* something which *is not*, is clearly a contradiction in terms."[2] He would presumably translate the sentence as $(\exists x)\sim(\exists y)$ $(y = x)$, so that if we read the quantifiers existentially, the sentence says that one and the same thing, i.e. $(y = x)$, does and does not exist.

In opposition to Quine, Küng and Canty have argued that Lesniewski was not interpreting quantification substitutionally (nor, for that matter, referentially, if 'reference' is understood narrowly in terms of the relation of singular denotation, i.e., naming).[3] Chief among their reasons for denying that Lesniewskian Quantification is substitutional is that an adequate account of his quantification in ontology must take account of the 'extensions' nouns can have, not merely the possible substituends for noun variables.

Of course, if there are n objects in the domain of discourse, there are 2^n extensions to be assigned to names. And if the universe of discourse is infinite (and Lesniewskian logic does not rule out this possibility), then there are more than denumerably many extensions. Now in this last case, there can be no system which contains or will contain constants for all the extensions, because any given system can contain only at most denumerably many definitions and axioms. We must distinguish between the false claim that more than denumerably many different constants can be introduced into the system and the true claim that each of the denumerably many constants which can be introduced can have any one of the more than denumerably many possible extensions.[4]

To answer the question of whether Lesniewskian quantification is substitutional, we must clarify exactly what is and what is not involved in the substitutional interpretation. The substitutional account given for a particular generalization is that the quantification is true if at least one of its substitution instances is true (a universal generalization is true if every instance is true). I noted earlier that the analysans segment of such a truth condition makes essential use of the notion of a substitution instance and not necessarily of notions such as that of a domain, values of variables, or sequences of objects satisfying an open sentence. If substitutional quantification is so construed, then non-substitutional quantification can be distinguished from it as not involving the concept of a substitution instance in accounting for the truth of a generalization. Tarski's account is the paradigm for non-substitutional quantification. He takes sequences as the objects satisfying a sentential function. Tarski says a sequence f satisfies an open sentence of the form of a universal quantification if and only if, no matter how we vary the sequence with respect to the variable bound by the quantifier, the open sentence is still satisfied. To repeat, the major difference between substitutional and non-substitutional quantification as presently explicated is that the former makes essential reference to substitution instances and the latter does not.

Another point made earlier was that the matter of how the instances of substitutional quantifications are themselves made true is independent of (and need not be given in) the analysans of the quantificational sentences. Consider the following truth condition intended to explicate substitutional quantification for a standard first order language:

> the valuation of $((x)A) = T$ iff the valuation of $((t/x)A = T$, for all singular expressions t.

Such a condition for a quantification is independent of the truth conditions for its instances, the '$((t/x)A)$' in the above. (The point is a simple one, and analogous to the case of truth conditions for truth functional sentences, such as conjunctions. These truth conditions don't mention how the conjuncts get their truth values.) To illustrate this further, let us contrast the views of two substitutional theorists. Our first theorist assigns truth values to instances in the usual way, e.g., 'Socrates is human' is true iff 'Socrates' refers to (is assigned) an individual that is also referred to by (is a member of the set assigned to) 'is human'. The resulting particular generalization '$(\exists x)(x$ is human)', (though explained substitutionally) does derivatively have a referential effect. For the second theorist the situation is quite different. She regards the instance 'Pegasus is a flying horse' as true even though no referents are assigned to its parts. The resulting generalization, taken substitutionally '$(\exists x)(x$ is a flying horse)', is true but need not have any referential function.

The present recommendation for the use of the locutions "substitutional" and "referential" quantification helps us to see that a generalization can be substitutional as well as referential (our first theorist);

or substitutional but not referential (our second theorist); or objectual but not substitutional (Tarski and Quine). Quine's opposition of substitutional and referential quantification is misleading. A quantification is interpreted substitutionally when its truth conditions essentially involve the notion of an instance and this does not preclude the possibility that the truth conditions for these instances secure the reference in question.

Now if we take seriously the independence of spelling out the truth conditions for instances from those for quantifications, Lesniewski's systems lend themselves very nicely to being interpreted as uniformly having substitutional truth conditions for quantifying with regard to diverse grammatical categories. That is to say, a universal generalization is true if all the instances are, where the instances are constructed by putting the appropriate substituends: sentences, sentence connectives, nouns, etc., for the appropriate variables. Of especial importance for elementary ontology are nouns as substituends, e.g., as in the definition of 'exists'

$(b) [\text{ex } b \equiv (\exists a) (a \text{ est } b)]$.

This leaves open, as it should, the question of how the instances become true. Lesniewski has answered this for the above. The substituends for variables like 'a' and 'b' are nouns which, depending on whether they are proper nouns, common nouns, empty nouns or universal nouns, are assigned one object as a referent, several or no objects as referents, or every object. The truth condition for crucial instances of the form '——est——' is that the subject noun be a proper noun and its referent be identical with a referent of the predicate noun. If viewed in this way Lesniewskian generalizations of the above sort, e.g. the definition of existence, are both substitutional and referential. A generalization such as '$(\exists b)$ (Socrates est b)' is true if it has at least one true substitution instance. Furthermore, it is because of the truth of such an instance that the generalization succeeds in referring.

Küng and Canty say that Lesniewskian quantification is not referential because they accept aspects of Quine's account of reference and quantification, viz., that substituends always be of the category of names and that naming or its surrogate (being the value of a variable from the grammatical category of names) is the only means for achieving reference. Quine's view of variables as being only of the category of singular terms was questioned in this monograph. More relevant here is the merely terminological point that 'refers' can be used equally well for all nouns and not just proper nouns (singular terms). If our terminology is correct, then we are justified in calling some Lesniewskian quantifications, such as those in elementry ontology, "referential."

How well the above reconstruction of parts of elementary ontology fits with what we know of Lesniewski is partly a historical question. It seems to be indirectly confirmed by looking at remarks by Kotarbinski and Ajdukiewicz, who were writing at the time.[5] Michael Dummett has recently provided the same sort of reconstruction for Lejewski's variant of Lesniewski's approach.[6]

A further question remains outstanding though—whether Lesniewskian quantification for other grammatical categories such as those found in prototethic or higher-order ontology is referential (in the above sense) as well as substitutional. That is, granted, for instance, that the sentence variables of prototethics have sentences as substituends, do these sentences as such refer, i.e., have referents assigned to them? Similarly, do the functor-predicates of ontology, e.g. 'ex', which can serve as substituends, have referents assigned to them? The answer is that if "refers" is used as it is customarily, then it appears to be restricted to nouns.

C.   In What Sense Is Lesniewski's Logic a Free Logic?

When Karel Lambert introduced the felicitous expression 'free logic' he intended by it two things: (1) allowing for vacuous terms, e.g., 'Pegasus', in logic; and (2) prohibiting certain sentences, such as those making existence claims, from being logical truths or theorems of logic. Ontology is a free logic in the first sense, since empty (vacuous) nouns are as allowable as any other nouns. It is the second sense of free logic, concerning logics without existence assumptions, which will occupy our attention. Lambert introduced his term to refer to systems such as those derivative from *Principia Mathematica* in which the notions of existence and quantification coincide. To ask *simpliciter* whether

Lesniewski's systems of logic are free in Lambert's second sense is to ask a confused question. In its place let us put the following three questions. (1) Are existence sentences (recall that for Lesniewski this is not merely a matter of looking to the particular quantifications) theorems in Lesniewski's systems? (2) Are any particular generalizations, assuming they involve reference, theorems? (3) Are any particular generalizations, even those which don't have referential force, theorems? The rationale for having a free logic (letting this term now cover all three of the above issues) is the desired topic neutrality of the subject. On this basis Russell recoiled at the prospect that a system of logic should resemble Anselm's ontological argument in establishing on purely logical grounds that something exists. In general, then, does the use solely of logical principles beg any of the three questions above of whether anything exists, whether the domain associated with one's quantifiers has members, and whether the class of substitution instances associated with one's quantifiers has members?

The answer to the first of these questions is that sentences of the form 'ex ——' are not theorems in Lesniewski's systems. In this sense, his systems of logic are free of existence assumptions. This point applies trivially to protothetic, which has quantifiers, but which in Lesniewski's formulation, does not contain the expression 'ex'.

The two remaining questions must be clearly distinguished. Quantification may be referential, by being construed non-substitutionally, as in Tarski, or by being construed substitutionally where the revelant substitution instances are assigned referents, as in our first theorist. Non-substitutional quantification is *per se* referential, whereas substitutional quantification is, if referential, only so *per accidens*. The set of the referents to which such quantifications refer is customarily called the domain of the quantifiers. A particular quantification having referential force will be true only if there is at least one member of the domain. To avoid prejudging this issue of domain membership we require that the logic of such quantifications should not include among its theorems any particular generalizations.

Now, what about quantification interpreted substitutionally but not referentially? Recall that possible candidates for such quantifications would be of those grammatical categories not normally considered to be referring expressions, e.g., sentences, sentence connectives, or of a category of which some constants do refer and some do not, e.g.,

'Socrates', 'Pegasus'. The truth conditions for such a particular generalization indicate that it would be false if there were no substitution instances whatever for the grammatical category in question.

Lejewski has presented an answer to our last two questions. His position is that for Lesniewski, particular generalizations can be theorems of logic, and as such they will be true in the empty domain, but would not be true if there were no relevant substitution instances. Lesniewskian quantification is free in the sense that it makes no requirements that its domain be non-empty, but not free in that it requires that there be substitution instances.

> Under the unrestricted interpretation, however, (16) $[(\exists x)$
> $(Fx \lor \sim Fx)]$ and (17) $[(x)Fx \supset (\exists x)(Fx)]$ come to be true
> irrespective of whether the universe is empty or non-empty. For (16)
> is implied by any component of type '$Fa \lor \sim Fa$' where 'a' stands for
> a noun-expression. In particular it is implied by a component 'Fa
> $\lor \sim Fa$' in which 'a' stands for an empty noun expression. Such a
> component is true for all choices of universe and so is (16). In the
> case of (17) we argue as follows: if we assume that the antecedent
> of (17) is true then a proposition of type 'Fa' where 'a' stands for an
> empty noun-expression must also be true in harmony with the un-
> restricted interpretation of the universal quantifier. Now any such
> proposition implies the proposition of type '$(\exists x)(Fx)$', which
> again must be true. Thus in the establishing of the truth value of
> (16) and (17) the problem of whether the universe is empty or
> non-empty is altogether irrelevant on condition, of course, that
> we adopt the unrestricted interpretation of the quantifiers.[7]

Lesniewskian quantification can plausibly be interpreted as being both substitutional and referential, and as a system of logic free of any assumptions about domain membership, i.e. referential, commitments but not free of certain substitutional commitments. This accords with our own earlier reading of Lesniewski. That Lesniewski's substitutional quantification is not free of the assumption that there are relevant substitution instances should come as no surprise. It follows from the constructive way in which he introduced quantifiers for non-primitive grammatical categories. A quantifier for a new category is adopted only after a constant of that category has been properly defined. Thus a quantifier for a monadic predicate in ontology attends upon the introduction of a constant of that type, e.g., $(\exists F)(fb)$ could be introduced after the predicate constant 'ex' was introduced by the definition ex $b \equiv (\exists a)(a \text{ est } b)$. Lesniewski's non-primitive quantifiers are introduced in what might be called a catergorially constructive fashion; we have a quantifier for a category only after constructing a constant-substituend of that category. This constraint on introducing quantifiers guarantees that there will always be instances to serve as the basis for substitutional generalizations.

Notes

1. These thoughts on the plausibility of a completeness proof for ontology were supplied by Jack C. Boudreaux. He has shown in a paper to appear in the *Notre Dame Journal of Formal Logic* that if one extends the "growing" language of ontology to a language like those of first-order logic (or for that matter to certain higher-order, type theoretic languages), then a Henkin type completeness proof is forthcoming.

2. W. V. O. Quine, *Mathematical Logic* (New York: Harper Torchbooks, 1951), p. 150.

3. G. Küng and J. T. Canty, "Substitutional Quantification and Lesniewskian Quantifiers," *Theoria*, 36, (1970), p. 179.

4. Ibid., pp. 179, 180-181.

5. T. Kotarbinski, *Gnosiology: The Scientific Approach to the Theory of Knowledge* (New York: Pergamon Press, 1966), pp. 132, 134-135, 530. Included in this translation is a 1930 review by K. Ajdukiewicz.

6. S. Körner, ed., *Philosophy of Logic* (Berkeley: University of California Press, 1976), pp. 28-42.

7. C. Lejewski, "Logic and Existence," in *Logic and Philosophy*, ed. by G. Iseminger (New York: Appleton-Century-Crofts, 1968), pp. 176-177.

Bibliography

Ajdukiewicz, K. "Die syntaktische Konnexität," *Studia Philosophica* (Lwow), 1 (1935), 1-27. (Translated and mimeographed as *Syntactic Connection* at the College of the University of Chicago in 1951. Reprinted in *Polish Logic 1920-1939*. Edited by S. McCall. Oxford: Clarendon Press, 1967.)

———. "On the Notion of Existence," *Studia Philosophica* (Poznan), 4 (1951), 8-22.

———. "From the Methodology of the Deductive Sciences," *Studia Logica*, 19 (1966), 9-45. (Translated by J. Giedymin.)

Ashworth, E. J. *Language and Logic in the Post-Medieval Period.* Dordrecht: D. Reidel, 1974.

———. "Existential Assumptions in Late Medieval Logic," *American Philosophical Quarterly* 10 (1973), 141-147.

Ayer, A. J. "Symposium: On What There Is," *Proceedings of the Aristotelian Society*, Suppl. 25 (1951).

Baier, K. "Existence," *Proceedings of the Aristotelian Society*, 41 (October, 1960), 19-40.

Bar-Hillel, Y. " On Mr. Sorensen's Analysis of 'To Be' and 'To Be True' "*Analysis*, 20 (March, 1960), 93-96.

Barrett, W. *Irrational Man*. Garden City: Anchor Books-Doubleday, 1962. (Appendix II.)

———. "One Hundred Real Dollars (or Doorknobs)," *Journal of Philosophy*, 59 (November, 1962), 763.

Beatty, R. "Peirce's Developement of Quantifiers and of Predicate

Logic," *Notre Dame Journal of Formal Logic*, 10 (January, 1969), 64-76.

Belnap, N. "Review of Hintikka's Existential Presuppositions and Existential Commitments," *Journal of Symbolic Logic*, 25 (1960), 88.

Benacerraf, P., and H. Putnam, eds. *Philosophy of Mathematics, Selected Readings*. Englewood Cliffs, N. J.: Prentice-Hall, 1964.

Bernays, P. "Review of Gödel's 'Russell's Mathematical Logic," *Journal of Symbolic Logic*, 11 (September, 1946), 75-79.

Beth, W. E. *The Foundations of Mathematics*. New York: Harper Torchbooks, 1966.

———. "Remarks on the Paradoxes of Logic and Set Theory." *Essays on the Foundation of Mathematics Dedicated to A. Frankel on His Seventieth Anniversary*. Edited by Y. Bar-Hillel. Jerusalem: Magnes Press, 1961.

Binkley, R. "A Note on Sorensen and Existence," *Analysis*, 20 (December, 1959), 48.

———. "Quantifying, Quotation, and a Paradox," *Nous*, 4 (September, 1970), 271-277.

Black, M. *The Nature of Mathematics*. Paterson, N. J.: Littlefield, Adams and Co., 1959.

Black, M. , A. Kapp, and N. Cooper "Report on Analysis Problem Number Three, 'Does the Logical Truth $(Ex)(Fx \text{ v} \sim Fx)$ Entail that at least One Individual Exists?', "*Analysis*, 14 (October, 1953), 1-5.

Bochenski, I. M. *A History of Formal Logic*. Notre Dame: University Press, 1961. (Translated by Ivo Thomas.)

———. *Logisch-philosophische Studien*. Edited by A. Menne. Notre Dame, Ind.: Freiburg, 1959. (English edition: Dordrecht, 1962.)

———. "On Syntactical Categories," *The New Scholasticism*, 23 (1949), 257-280.

Bochenski, J. M., A. Church, and N. Goodman. *The Problem of Universals. A Symposium*. Notre Dame, Ind.: Freiburg, 1956.

Boehner, P. *Medieval Logic (1250-1400)*. Manchester: University Press, 1952.

Canty, J. T. "The Numerical Epsilon," *Notre Dame Journal of Formal Logic*, 10 (1969), 47-63.

————. "Ontology: Lesniewski's Logical Language," *Foundations of Language*, 5 (1969), 455-469.

————. "Elementary Logic Without Referential Quantification," 1970. (Mimeographed.)

Carnap, R. *Logische Syntax der Sprache*. Vienna: 1934 (Translation with additions: *The Logical Syntax of Language*. Paterson, N. J.: Littlefield, Adams and Co., 1959.)

————. *Introduction to Semantics and the Formalization of Logic*. Harvard: University Press, 1943.

————. *Meaning and Necessity*. Chicago: University of Chicago Press, 1947. (2nd ed., 1956.)

————. "The Elimination of Metaphysics Through the Logical Analysis of Language." *Logical Positivism*. Edited by A. J. Ayer. Glencoe, Ill.: The Free Press, 1959.

Cartwright, R. L. "Ontology and the Theory of Meaning," *Philosophy of Science*, 21 (1954), 316-325.

————. "Negative Existentials," *Journal of Philosophy*, 57 (1960), 629-640.

Caton, C. E. (ed.) *Philosophy and Ordinary Language*. Urbana: University of Illinois Press, 1963.

Child, J., and F. I. Goldberg. " 'Exists' as a Predicate: A Reconsideration," *Analysis*, 31 (December, 1970), 53-57.

Chisholm, R. "Beyond Being and Nonbeing," *New Readings in Philosophical Analysis*. Edited by H. Feigl, W. Sellars, and K. Lehrer. New York: Appleton-Century-Crofts, 1972.

Church, A. *Introduction to Mathematical Logic*. Vol. I. New Jersey: Princeton University Press, 1956.

————. "Ontological commitment," *The Journal of Philosophy*, 55 (1958), 1008-1014.

————. "Review of Leonard, 'The Logic of Existence,' " *The Journal of Symbolic Logic*, 28 (1963), 259 ff.

————. "A History of the Question of Existential Import of Categorical Propositions," *Logic, Methodology and Philosophy of Science, Proceedings of the 1967 International Congress*. Edited by B. van Rootselaar and J. F. Staal. Amsterdam: North Holland Publishing Co., 1968.

————. "Review of Lambert, 'Existential Import Revisted,' " *The*

Journal of Symbolic Logic, 30 (1965), 101-102.

Cocchiarella, N. B. "Some Remarks on Second Order Logic With Existence Attributes," *Nous*, 2 (May, 1968), 165-175.

———. "Second Order Logic With Existence Attributes," *The Journal of Symbolic Logic*, 34 (March, 1969).

———. "Existence Entailing Attributes, Modes of Copulation and Modes of Being in Second Order Logic," *Nous*, 3 (February, 1969), 33-48.

Cohen, J. *The Diversity of Meaning*. London: Methuen and Co., 1962.

Copi, I., and A. J. Gould. *Contemporary Readings in Logical Theory*. New York: Macmillan Co., 1967.

Danto, A. C. *Analytical Philosophy of Knowledge*. Cambridge: Cambridge University Press, 1968.

Donagan, A. "Recent Criticisms of Russell's Analysis of Existence," *Analysis*, 12 (1952), 132-137.

Dryer, D. P. "The Concept of Existence in Kant," *The Monist*, 50 (1966), 17-33.

Dummett, M. *Frege*. London: Duckworth, 1973.

———. "Comment" (on Lejewski's "Ontology and Logic"), *Philosophy of Logic*. Edited by S. Körner. Los Angeles: University of California Press, 1976.

Dunn, M. and N. D. Belnap, Jr., "The Substitution Interpretation of the Quantifiers," *Nous*, 2 (1968), 177-185.

Ebersole, F. B. "Whether Existence Is a Predicate," *The Journal of Philosophy*, 60 (August, 1963), 509-523.

Feigl, H. and W. Sellars, eds. *Readings in Philosophical Analysis*. New York: Appleton-Century-Crofts, 1949.

Fine, A. L. "Quantification Over the Real Numbers," *Philosophical Studies*, 19 (January-February, 1969), 27-32.

Follesdal, D. "Knowledge, Identity and Existence," *Theoria*, 33 (1967), 1-27.

———. "Interpretation of Quantifiers." *Logic, Methodology and Philosophy of Science, Proceedings of the 1967 International Congress.* Edited by B. van Rootselaar and J. F. Staal. Amsterdam: North Holland Publishing Co., 1968.

Frege, G. "Begriffschrift." Translated by S. Bauer-Mengelberg in *From Frege to Gödel, A Source Book in Mathematical Logic 1879-1931*. Edited by J. van Heijenoort. Cambridge: Harvard University Press, 1967.

————. *The Foundations of Arithmetic.* Translated by J. L. Austin. New York: Philosophical Library, 1950.

————. *Translations from the Philosophical Writings of Gottlob Frege.* Edited by P. Geach and M. Black. Oxford: Basil Blackwell, 1960.

————. *Nachgelassene Schriften.* Vol. I. Hamburg: Felix Meiner Verlag, 1969.

————. "On Sense and Nominatum." *Readings in Philosophical Analysis.* Edited by H. Feigl and W. Sellars. Appleton-Century-Crofts, New York, 1949.

Geach, P. "Russell's Theory of Descriptions," *Analysis*, 10 1950, 84-88.

————. "Symposium: On What There Is," *Proceedings of the Aristotelian Society*, Suppl. 25 (1951).

————. "Form and Existence," *Proceedings of the Aristotelian Society*, Suppl. 28 (1954-1955).

————. "Assertion," *Philosophical Review*, 74 (1965), 449-465.

————. *Logic Matters.* Los Angeles: University of California Press, 1972.

Geach, P. and R. H. Stoothoff. "Symposium: What Actually Exists," *Proceedings of the Aristotelian Society*, Suppl. 42 (1968).

Gilson, E. *Being and Some Philosophers.* Toronto: Pontifical Institute of Medieval Studies, 1949.

Goe, G. "A Reconstruction of Formal Logic," *Notre Dame Journal of Formal Logic*, 7 (1966), 129-158.

Gottlieb, D. "A Method for Ontology, with Applications to Numbers and Events," *Journal of Philosophy*, 73 (1976).

Graham, A. C. " Being' in Classical Chinese," *The Verb "Be" and Its Synonyms.* Vol. I. Edited by J. W. M. Verhaar. New York: Humanities Press, 1966.

Grzegorczyk, A. *An Outline of Mathematical Logic.* Dordrecht, Holland: D. Reidel, 1974.

————. "The Systems of Lesniewski in Relation to Contemporary Logical Research," *Studia Logica*, 3 (1955), 77-95.

Hailperin, T. "Quantification Theory and Empty Individual Domains," *Journal of Symbolic Logic*, 18 (1953), 197-200.

Hailperin, T., and H. Leblanc. "Nondesignating Singular Terms," *Philosophical Review*, 68 (1959), 129-136.

Harman, G. "Quine on Meaning and Existence, I and II," *Review of Metaphysics*, 21 (September, 1967), 124-151; (December, 1967), 343-367.

———. "Substitutional Quantification and Quotation," *Nous*, 5 (May, 1971), 213-214.

Henkin, L. "Some Notes on Nominalism," *Journal of Symbolic Logic*, 18 (March, 1953), 19-29.

Hick, J. and A. G. McGill. *The Many Faced Argument*. New York: Macmillan, 1967.

Hilbert, D. and W. Ackermann. *Principles of Mathematical Logic*. New York: Chelsea Publishing Co., 1950.

Hintikka, J. *Models for Modalities*. New York: Humanities Press, 1969.

———. "Existential Presuppositions and Existential Commitments," *Journal of Philosophy*, 56 (1959), 125-137.

———. "Towards a Theory of Definite Descriptions," *Analysis*, 19 (1959), 79-85.

———. "Definite Descriptions and Self-identity," *Philosophical Studies*, 15 (1964), 5-7.

———. "Language Games for Quantifiers," American *Philosophical Quarterly, Monograph Series*, 2 (1968), 46-72.

———. "The Semantics of Modal Notions and the Indeterminacy of Ontology," *Synthese*, 21 (1970), 408-424.

Hiz, H. "Kotarbinski on Truth," University of Pennsylvania Transformation and Discourse Analysis Papers, 1966. (Mimeographed.)

———. "The Aletheic Semantic Theory," *The Philosophical Forum*, 1 (1969), 438-451.

———. "Referentials," *Semiotica*, 1 (1969), 136-166.

———. Fragments from "On the Assertions of Existence," 1971. (Mimeographed.)

Hook, S. *The Quest for Being*. New York: St. Martin's Press, 1961.

Hughes, G. E. and M. J. Cresswell. *An Introduction to Modal Logic*. London: Methuen and Co., 1968.

Iseminger, G., ed. *Logic and Philosophy*. New York: Appleton-Century-Crofts, 1968.

Jasowski, S. "On the Rules of Suppositions in Formal Logic," *Studia Logica*, 1 (1934). (Reprinted in McCall.)

Johnson, W. E. *Logic*. Vol. 1. New York: Dover Publications, Inc., 1964.

Joseph, H. W. B. *An Introduction to Logic*. Oxford: Oxford University Press, 1966.

Kahn, C. H. "The Greek Verb 'To Be' and the Concept of Being," *Foundations of Language*, 2 (August, 1966), 245-265.

Kalish, D. "Mr. Pap on Logic, Existence and Descriptions," *Analysis*, 15 (1954), 61-65.

Kalish, D., and R. Montague. *Logic: Techniques of Formal Reasoning*. New York: Harcourt Brace & Co., 1964.

Kant, I. *Critique of Pure Reason*. Translated by N. K. Smith. London: Macmillan and Co., 1953.

Kaplan, D. "What is Russell's Theory of Descriptions?" *Physics, Logic and History*. Edited by W. Yourgrau. New York: Plenum Press, 1970.

———. "Bob and Carol and Ted and Alice," *Approaches to Natural Language*. Edited by J. Hintikka, J. M. E. Moravcsik, and P. Suppes. Dordrecht: D. Reidel, 1973.

Kearns, J. T. "The Contribution of Lesniewski," *Notre Dame Journal of Formal Logic*, 8 (April, 1967), 61-93.

———. "The Logical Concept of Existence," *Notre Dame Journal of Formal Logic*, 9 (October, 1968), 313-324.

———. "Two Views of Variables," *Notre Dame Journal of Formal Logic*, 10 (April, 1969), 163-180.

Keynes, J. M. *Studies and Exercises in Formal Logic*. 4th ed. New York: Macmillan Co., 1906.

Kiteley, M. "Is Existence a Predicate?" *Mind*, 73 (1964), 364-373.

Kneale, W. and M. *The Development of Logic*. Oxford: The Clarendon Press, 1962.

Körner, S. *Philosophy of Logic*. Los Angeles: University of California Press, 1976.

Kotarbinski, T. *Elementy teorii poznania, logiki formalnej i metodoligii nauk*. Lwow. 2nd ed. Warsaw: Ossoleum, 1961.

———. *Leçons Sur L'Histoire de la Logique*. Translated by A. Posner. Warsaw: Polish Scientific, 1965.

———. *Gnosiology*. Translated by O. Wojtasiewicz. London: Pergamon Press, 1966.

Kripke, S. "Is There a Problem about Substitutional Quantification?" *Truth and Meaning*. Edited by G. Evans and J. McDowell. Oxford: Oxford University Press, 1976.

Küng, G. *Ontology and the Logistic Analysis of Language*. New York: Humanities Press, 1967.

Küng, G. and J. T. Canty, "*Substitutional Quantification and Lesniewskian Quantifiers*," *Theoria* (36), 1970, 165-182.

Lambert, K., ed. *The Logical Way of Doing Things*. New Haven: Yale University Press, 1969.

————. *Philosophical Problems in Logic.* Holland: Dordrecht, 1970.

————. "Notes on E!," *Philosophical Studies,* 9 (1958), 60-64.

————. "Notes on E! II," *Philosophical Studies,* 12 (1961), 1-5.

————. "Notes on E! III," *Philosophical Studies,* 13 (1962), 51-58.

————. "Existential Import Revisited," *Notre Dame Journal for Formal Logic,* 4 (1963), 133-144.

————. "Notes in E! IV: a Reduction in Free Quantification Theory with Identity and Descriptions," *Philosophical Studies,* 15 (1964), 85-88.

————. "On Logic and Existence," *Notre Dame Journal for Formal Logic,* 6 (1965), 135-141.

————. "Quantification and Existence," *Inquiry,* 6 (Winter, 1963), 319-324.

Lambert, K. and B. C. van Fraasen. *Derivation and Counterexample.* Belmont: Dickenson Publishing Co., 1972.

Leblanc, H. *Truth-Value Semantics.* Amsterdam: North Holland Publishing Co., 1976.

————. ed. Truth, Syntax and Modality. Amsterdam: North Holland Publishing Co., 1973.

————. "A Simplified Account of Validity and Implication for Quantificational Logic," *Journal of Symbolic Logic,* 33 (1968), 231-235.

————. "On Meyer and Lambert's Quantificational Calculus FQ," *Journal of Symbolic Logic,* 33 (1968), 275-280.

————. "A Simplified Strong Completeness Proof for QC," forthcoming in Akten des XIV. Internationalen Kongresses für Philosophie, Wien: 2.-9. September, 1968.

————. "Truth-value Semantics for a Logic of Existence," *Notre Dame Journal of Formal Logic,* 12 (April, 1971).

————. "On Dispensing With Things and Worlds," Paper given at colloquium on Ontology and Existence, New York University, Spring, 1971, to be published by New York University Press.

Leblanc, H. and T. Hailperin. "Nondesignating Singular Terms," *The Philosophical Review,* 68 (1959), 129-136.

Leblanc, H. and R. K. Meyer. "Open Formulas and the Empty Domain," *Archiv für Mathematische Logik und Grundlagenforschung,* 12 (1969), 78-84.

Leblanc, H. and R. H. Thomason. "Completeness Theorems for Some Presupposition-free Logics," *Fundamenta Mathematicae,* 62 (1968), 125-164.

Lejewski, C. "Logic and Existence," *British Journal for the Philosophy of Science*, 5 (1954), 104-119.

———. "Logic and Existence," Cited in *Logic and Philosophy*. Edited by G. Iseminger. New York: Appleton-Century-Crofts, 1968.

———. "Proper Names," *Proceedings of the Aristotelian Society*, Suppl. 31 (1957), 229-256.

———. "On Lesniewski's Ontology," *Ratio*, 1 (1958), 150-176.

———. "A Re-examination of the Russellian Theory of Descriptions," *Philosophy*, 35 (1960), 14-29.

———. "Aristotle's Syllogistic and Its Extensions," *Synthese*, 15 (1963), 125-154.

———. "The Problem of Ontological Commitment," *Fragmenty Filozoficzne Seria Trzecia Ksiega Pamiatkowa Ku Czi Profesora Tadeusza Kotarbinskiego*. Warsaw: Panstwowe Wydawnictwo Naukowe, 1967.

———. "Quantification and Ontological Commitment." *Physics, Logic and History*. Edited by W. Yourgrau. New York: Plenum Press, 1970.

———. "Ontology and Logic," *Philosophy of Logic*. Edited by S. Körner. Los Angeles: University of California Press, 1976.

Leonard, H. S. "The logic of Existence," *Philosophical Studies*, 7 (June, 1956), 49-64.

———. "Essences, Attributes and Predicates." Presidential Address for the sixty-second annual meeting of the Western Division of the American Philosophical Association, Milwaukee, Wisconsin, April-May, 1964.

Lesniewski, S. "Über die Grundlager der Ontologie, " *Comptes rendus des séances de la Société des Sciences et des Lettres de Varsovie*, Cl. 3, 23 (1930), 111-132.

———. "Über Definitionen in der sogenannten Theorie der Deduktion," *Comptes rendus des séances de la Société des Sciences et des Lettres de Varsovie*, Cl. 3, 24 (1931), 289-309. (Translated and reprinted in *Polish Logic 1920-1939*. Edited by S. McCall. Oxford: Clarendon Press, 1967.)

———. "Einleitende Bermerkungen zur Fortsetzung meiner Mitteilung u. d.t. 'Grundzüge eines neuen Systems der Grundlagen der Mathematik.'" Offprint of the never published *Collectanea Logica* (Warszawa). (Translated and reprinted in *Polish Logic 1920-1939*. Edited by S. McCall. Oxford: Clarendon Press, 1967.)

———. "Grundzüge eines neuen Systems der Grundlagen der Mathematik 12." Offprint of the never published *Collectanea Logica* (Warszawa), 1938. (Copies in Harvard and Münster, Westfalen.)

Linsky, L. *Semantics and the Philosophy of Language*. Urbana: University of Illinois Press, 1952.

———. *Referring*. New York: Humanities Press, 1967.

———, ed. *Reference and Modality*. Oxford: Oxford University Press, 1971.

Lukasiewicz, J. *Aristotle's Syllogistic From the Standpoint of Modern Formal Logic*. Oxford: Oxford University Press, 1951.

———. *Elements of Mathematical Logic*. Translated by O. Wojtasiewicz. London: Pergamon Press, 1963.

Lukasiewicz, J., E. Anscombe, and K. Popper. "Symposium." *Proceedings of the Aristotelian Society*, Suppl. 27 (1953), 68-120. ("The Principle of Individuation.")

Luschei, E. C. *The Logical Systems of Lesniewski*. Amsterdam: North Holland Publishing Co., 1962.

Lyons, J. "Existence, Location, Possession and Transitivity." *Logic, Methodology and Philosophy of Science. Proceedings of the 1967 International Congress*. Edited by B. van Rootselaar and J. F. Staal. Amsterdam: North Holland Publishing Co., 1968.

Marcus. R. B. "Interpreting Quantification," *Inquiry*, 5 (1962), 252-259.

———. "Reply to Dr. Lambert," *Inquiry*, 6 (Winter, 1963), 325-327.

———. "Modal Logics: I, Modalities and Intensional Languages." Boston Studies in the Philosophy of Science. *Proceedings of the Colloquium of 1961/62*. Dordrecht (1963), 97-104.

———. "Modal Logic." *Contemporary Philosophy*. Edited by R. Klibansky. Florence: La Nuova Italia Editrice, 1968.

———. "Dispensing with Possibilia," *Proceedings and Addresses of the American Philosophical Association*. Vol. XLIX, Nov. 1976. University of Delaware, Newark: The American Philosophical Association, 1976.

Margolis, J. *Knowledge and Existence*. New York: Oxford University Press, 1973.

———, ed. *Fact and Existence*. Toronto: University of Toronto Press, 1969.

Martin, R. M. *Truth and Denotation*. London: Routledge and Kegan Paul, 1958.

———. *Belief, Existence and Meaning.* New York: New York University Press, 1969.

———. *Logic, Language and Metaphysics.* New York: New York University Press, 1971.

———. "A Homogeneous System for Formal Logic," *The Journal of Symbolic Logic*, 8 (March, 1943), 1-59.

———. "Truth and Multiple Denotation," *The Journal of Symbolic Logic*, 18 (March, 1953), 11-18.

———. "On Non-Translational Semantics," *Proceedings of the XIth International Congress of Philosophy*, 5 (1953), 132-138.

———. "On Denotation and Ontic Commitment," *Philosophical Studies*, 13 (April, 1962), 35-39.

———. "Existential Quantification and the 'Regimentation' of Ordinary Language," *Mind*, 71 (October, 1962), 525-529.

———. "On Ontology and the Province of Logic: Some Critical Remarks," *Essays in Honor of I. M. Bochenski.* Edited by A Tymeniecka. Amsterdam: North Holland Publishing Co., 1965.

———. "Of Time and the Null Individual," *Journal of Philosophy*, 62 (December, 1965), 723-735.

———. "On Non-Translational and Aletheic Semantic Theory," 1971. (mimeographed.)

Mates, B. "Leibniz on Possible Worlds." *Logic, Methodology and Philosophy of Science. Proceedings of the 1967 International Congress.* Amsterdam: North Holland Publishing Co., 1968.

McCall, S., ed. *Polish Logic 1920-1939.* Oxford: Clarendon Press, 1967.

Meyer, R. K. and K. Lambert. "Universally Free Logic and Standard Quantification Theory," *The Journal of Symbolic Logic*, 33 (1968), 8-26.

Mill, J. S. *A System of Logic.* London: Longmans, Green and Co., 1961.

Moody, E. A. *Truth and Consequence in Medieval Logic.* Amsterdam: North Holland Publishing Co., 1953.

Moore, G. E. *Philosophical Papers.* New York: Collier Books, 1959.

Moravcsik, J. M. E., ed. *Aristotle.* Garden City: Anchor Books-Doubleday, 1967.

———. "Being and Meaning in the *Sophist,*" *Acta Philosophica Fennica* 1962, 23-78.

Mostowski, A. *Logika Matematyczna*. Warsaw: Wroclaw, 1948.
————. "On the Rules of Proof in the Pure Functional Calculus of First Order," *Journal of Symbolic Logic*, 16 (1951), 107-111.
————. "On a Generalization of Quantifiers," *Fundamenta Mathematicae*, 44 (1957), 12-36.
Mostowski, A., and H. Rasiowa. "O Geometryccznej Interpretacji Wyrazen Logicznej," *Studia Logica*, 1 (1953), 254-275.
Munitz, M. K. *The Mystery of Existence*. New York: Appleton-Century-Crofts, 1965.
————. *Existence and Logic*. New York: New York University Press, 1974.
————, ed. *Logic and Ontology*. New York: New York University Press, 1973.
Nakhnikian, G., and W. Salmon, "Existence as a Predicate," *Philosophical Review*, 66 (1957), 535-542.
Owens, J. *An Interpretation of Existence*. Milwaukee: Bruce Publishing Co., 1968.
Pap, A. "Logic, Existence and The Theory of Descriptions," *Analysis*, 13 (1952), 97-111.
Parsons, C. "Ontology and Mathematics," *Philosophical Review*, 80 (April, 1971), 151-176.
————. "A Plea for Substitutional Quantification," *Journal of Philosophy*, 68 (April, 1971), 231-237.
Paton, H. J. *Kant's Metaphysics of Experience*. Vol. II. London: George Allen and Unwin Ltd., 1951.
Peirce, C. S. *Collected Papers*. Edited by C. Hartshorne and P. Weiss. Cambridge: Harvard University Press, 1931-1935.
Plantinga, A., ed. *The Ontological Argument*. Garden City: Anchor Books-Doubleday, 1965.
————. "Kant's Objection to the Ontological Argument," *Journal of Philosophy*, 63 (October, 1966), 537-545.
Prior, A. N. *Formal Logic*. Oxford: Oxford University Press, 1962.
————. *Objects of Thought*. Oxford: Oxford University Press, 1971.
————. *The Doctrine of Propositions and Terms*. Amherst: University of Massachusetts Press, 1976.
Quine, W. V. *A System of Logistic*. Cambridge: Harvard University Press, 1934.
————. *Mathematical Logic*. Rev. ed. New York: Harper Torchbooks, 1951.

———. *Elementary Logic*. Rev. ed. New York: Harper Torchbooks, 1965.

———. *Methods of Logic*. Rev. ed. Holt, Rinehart and Winston, 1963.

———. *Word and Object*. New York: Wiley and Sons, 1960.

———. *Set Theory and Its Logic*. Cambridge: Harvard University Press, 1963.

———. *Selected Logic Papers*. New York: Random House, 1966.

———. *The Ways of Paradox and Other Essays*. New York: Random House , 1966.

———. *Ontological Relativity and Other Essays*. New York: Columbia University Press, 1969.

———. *Philosophy of Logic*. Englewood Cliffs, N.J.: Prentice Hall, Inc., 1970.

———. *The Roots of Reference*. La Salle: Open Court, 1973.

———. "Designation and Existence," *Journal of Philosophy*, 36 (1939), 701-709. (Reprinted in *Readings in Philosophical Analysis*. Edited by H. Feigl and W. Sellars. New York: Appleton-Century-Crofts, 1949.)

———. "Notes on Existence and Necessity," *Journal of Philosophy*, 40 (1943), 113-127. (Reprinted in *Semantics and the Philosophy of Language*. Edited by L. Linsky. Urbana: University of Illinois Press, 1952.)

———. "On Universals," *Journal of Symbolic Logic*, 12 (1947), 74-84.

———. "Review of K. Ajdukiewicz, 'On the Notion of Existence,'" *Studia Philosophica*, 4 (1951), 7-22; in *Journal of Symbolic Logic*, 17 (1952), 141-142.

———. "Unification of Universes in Set Theory," *Journal of Symbolic Logic*. 21 (September, (1956), 267-279.

———. "Comment on the Paper of Marcus, R. B., 'Modal Logics I: Modalities and Intensional Languages,'" Boston Studies in the Philosophy of Science (1963), 97-104.

———. "Russell's Ontological Development," *Journal of Philosophy*, 63 (November, 1966), 657-667.

———. "Introductory Essays." *From Frege to Gödel; A Source Book in Mathematical Logic 1879-1931*. Edited by J. van Heijenoort. Cambridge: Harvard University Press, 1967.

———. "Existence." *Physics, Logic and History*. Edited by W. Yourgrau. New York: Plenum Press, 1970.

Ramsey, F. P. *The Foundations of Mathematics and Other Logical*

Essays. Paterson: Littlefield Adams and Co., 1960.

Raphael, D. D. "To Be and Not To Be," *Proceedings of the Aristotelian Society*, 61 (1960).

Reichenbach, H. *Elements of Symbolic Logic*. New York: Free Press of Glencoe, 1947.

Rescher, N. *Topics in Philosophical Logic*. New York: Humanities Press, 1968.

Rogers, R. "A Survey of Formal Semantics," *Synthese*, 15 (1963), 17-56.

Routley, R. "Some Things Do Not Exist," *Notre Dame Journal of Formal Logic*, 7 (July, 1966), 251-276.

———. "Existence and Identity in Quantified Modal Logics," *Notre Dame Journal of Formal Logic*, 10 (April, 1969), 113-149.

———. "Non-existence Does Not Exist," *Notre Dame Journal of Formal Logic*, 11 (July, 1970), 289-320.

Russell, B. *The Principles of Mathematics*. 1st ed. London: George Allen and Unwin, 1903.

———. *The Problems of Philosophy*. Oxford: Oxford University Press, 1912.

———. *Introduction to Mathematical Philosophy*. London: George Allen and Unwin Ltd., 1919.

———. *Logic and Knowledge*. New York: Macmillan, 1956.

———. *The Autobiography of Bertrand Russell 1872-1914*. New York: Bantam Books, 1968.

———. "On Some Difficulties in the Theory of Transfinite Numbers and Order Types," *Proceedings of the London Mathematical Society*, 4 (1906), 29-53.

———. "The Substitutional Theory of Classes and Relations." (This paper was communicated to the London Mathematical Society and mentioned though not published in their Proceedings for May 10, 1906. A Xerox copy of Russell's handwritten original is available from the Russell Archives, McMaster University, Canada.)

———. "Mr. Strawson on Referring," *Mind*, 66 (1957). (Reprinted in *Contemporary Readings In Logical Theory*. Edited by I. Copi and J. A. Gould. New York: Macmillan Co., 1967.)

Russell, B. and A. N. Whitehead. *Principia Mathematica*. Cambridge: Cambridge University Press, 1962.

Ryle, N. "Systematically Misleading Expressions." *Logic and Language*. Edited by A. Flew. Garden City: Anchor Books, 1965.

Scheffler, I. and N. Chomsky. "What Is Said To Be," *Proceedings of the Aristotelian Society*, 59 (1959), 71-82.

Schock, R. *Logics Without Existence Assumptions*. Stockholm: Almqvist and Wiksell, 1968.

Scott, D. "Existence and Description in Formal Logic." *Bertrand Russell: Philosopher of the Century*. Edited by R. Schoeman. London: Allen and Unwin, 1967.

Searle, J. R. *Speech Acts*. Cambridge: Cambridge University Press, 1969.

Sellars, W. E. *Science, Perception and Reality*. New York: Humanities Press, 1963.

Shearn, M. "Russell's Analysis of Existence," *Analysis*, 11 (1950), 124-131.

Slupecki, J. "St. Lesniewski's Protothetics," *Studia Logica*, 1 (1953), 44-111, 299.

–––. "St. Lesniewski's Calculus of Names," *Studia Logica*, 3 (1955), 7-76.

Smiley, T. "Sense Without Denotation," *Analysis*, 20 (1960), 125-135.

Sobocinski, B. "L'analyse de l'antinomie russellienne par Lesniewski," *Methodos*, 1 (1949), 94-107, 220-228, 308-316; 2 (1950), 237-257.

–––. "On Well Constructed Axiom Systems," *Rocznik Polskiego Towarzystwa Naukowego na Obczyznie*, 6 (1955-56), 54-65.

Sorensen, H. S. "An Analysis of 'To Be' and 'To Be True,'" *Analysis*, 19 (June, 1959), 121-131.

Strawson, P. F. *Introduction to Logical Theory*. London: Methuen and Co., Ltd., 1952.

–––. *Individuals*. New York: Doubleday, 1963.

–––, ed. *Philosophical Logic*. London: Oxford University Press, 1967.

–––. "On Referring" *Mind*, 59 (1950). (Reprinted in *Philosophy and Ordinary Language*. Edited by C. E. Caton. Urbana: University of Illinois Press, 1963.)

–––. "A Reply to Mr. Sellars," *Philosophical Review*, 63 (1954), 216-231.

–––. "A Logician's Landscape," *Philosophy*, 30 (1955), 229-237.

–––. "Is Existence Never a Predicate?," *Critica*, 1 (January, 1967), 5-15.

–––. "Positions for Quantifiers" *Semantics and Philosophy*. Edited by M. K. Munitz and P. Unger. New York: New York University Press, 1974.

Tarski, A. *Logic, Semantics, Metamathematics: Papers from 1923 to 1938.* Oxford: Oxford University Press, 1956. (Translated by J. H. Woodger.)

——. "The Semantic Conception of Truth," *Philosophy and Phenomenological Research* 4 (1944), 341-375. (Reprinted in *Semantics and the Philosophy of Language.* Edited by L. Linsky. Urbana: University of Illinois Press, 1952.)

van Fraassen, B. C. *Formal Semantics and Logic.* New York: Macmillan Co., 1971.

——. "Singular Terms, Truth-Value Gaps, and Free Logic," *Journal of Philosophy*, 63 (September, 1966), 481-495.

——. "The Completeness of Free Logic," *Zeitschrift für Mathematische Logik und Grundlagen der Mathematik,* 12 (1966), 219-234.

——. "Presupposition, Implication and Self-reference," *Journal of Philosophy*, 65 (1968), 136-151.

——. "Presuppositions, Supervaluations and Free Logic." *The Logical Way of Doing Things.* Edited by K. Lambert. New Haven: Yale University Press, 1969.

van Heijenoort, J., ed. *From Frege to Gödel: A Source Book in Mathematical Logic 1879-1931.* Cambridge: Harvard University Press, 1967.

van Rootselaar, B. and J. F. Staal. eds. *Logic, Methodology and Philosophy of Science. Proceedings of the 1967 International Congress.* Amsterdam: North Holland Publishing Co., 1968.

Verhaar, J. W. M., ed. *The Verb "Be" and Its Synonyms.* Vol. I and II. New York: Humanities Press, 1966; 1969.

Wallace, J. "On the Frame of Reference," *Synthese*, 22 (1970), 117-150.

——. "Convention T and Substitutional Quantification," *Nous,* 5 (May, 1971), 199-211.

Wang, H. "Process and Existence in Mathematics." *Essays On the Foundation of Mathematics, Dedicated to A. Frankel on His Seventieth Anniversary.* Edited by Y. Bar-Hillel. Jerusalem: Magnes Press, 1961.

Welker, D. "Existential Statements," *Journal of Philosophy*, 67 (June, 1970), 376-388.

White, M. *Toward Reunion in Philosophy.* Cambridge: Harvard University Press, 1958.

———. "Ontological Clarity and Semantic Obscurity," *Journal of Philosophy,* 48 (1951), 373-380.

Williams, D. C. "Dispensing With Existence, *Journal of Philosophy,* 59 (November, 1962), 748-762.

Wittgenstein, L. *Notebooks, 1914-1916.* Edited by G. H. von Wright and G. E. M. Anscombe. Translated by G. E. M. Anscombe. New York: Harper Torchbooks, 1961.

———. *Tractatus Logico-Philosophicus.* Translated by D. F. Pears and B. F. McGuiness. London: Routledge and Kegan Paul, 1961.

Wolff, R. P. *Kant's Theory of Mental Activity.* Cambridge: Harvard University Press, 1963.

———, ed. *Kant.* Garden City: Anchor Books-Doubleday, 1967.

Wu, J. S. "The Problem of Existential Import (From George Boole to P. F. Strawson)," *Norte Dame Journal of Formal Logic,* 10 (October, 1969), 415-425.

Yourgrau, W., ed. *Physics, Logic and History.* New York: Plenum Press, 1970.

Index